# SECRETS *of* SCIENCE

**Exciting, unbelievable,
incredible . . . and all true!**

SECRETS of SCIENCE

GRAHAM PHILLIPS

SUN

M

Copyright © Graham Phillips 1990

First published 1990 by Sun Books
THE MACMILLAN COMPANY OF AUSTRALIA PTY LTD
107 Moray Street, South Melbourne 3205
6 Clarke Street, Crows Nest 2065

Associated companies throughout the world

National Library of Australia
cataloguing in publication data

Phillips, Graham, 1958 –
    Secrets of science.
    Bibliography.
    ISBN 0 7251 0598 4.

    1. Science — Popular works. I. Title.

500

Set in Garamond Book by Savage Type Pty Ltd, Brisbane
Printed in Hong Kong

Illustrations by Christopher Haddon

# Contents

# *Author's note*

*I*n 1989 a surprising report appeared in the journal *Nature* (Volume 340, 6 July 1989). A survey conducted in both Britain and the United States revealed that an astonishingly high percentage of the public had a very limited knowledge of science. The majority of those interviewed, for example, did not know that the Earth goes around the Sun once a year or that antibiotics are ineffective against viruses.

Equally striking, however, was that the people interviewed claimed they were more interested in science than, for example, sport. Generally, you would expect people who are interested in a particular subject to also be well informed about it. This doesn't seem to be the case where science is concerned. For many people, much of science seems to remain secret.

My years as a scientist, when I spent a lot of time explaining my work to others, made it abundantly clear that when it comes to science there is a big communication problem. I think a large part of this problem is that there is a lack of science writing at the grass roots level — science written for the person who doesn't already have an understanding of and interest in the subject. This is what inspired me to write my newspaper columns.

Here is a collection of some of the articles I have written. It is a book I believe anyone can understand, and is presented in a way that will, I hope, keep the reader both entertained and informed.

# How to travel through time

*T*he day I learnt that time travel had been discovered by science remains one of my most vivid memories. I was only sixteen at the time and, like most sixteen-year-olds, a visit to the future was among my wildest dreams. Dreams, however, is all I thought they'd ever be. So you can imagine my excitement as I read about time travel, not in a science fiction novel, but in the pages of a textbook.

Time travel was discovered by Einstein with his Special Theory of Relativity. He showed that by travelling through space at very high speeds, you automatically pass forward in time. Microscopic time travellers, called muons, actually whizz through our atmosphere all the time. These small particles are created high up in the Earth's atmosphere as cosmic rays collide with nitrogen and oxygen atoms. The key to their time travel is not a time machine, it's purely speed. Because they travel at very high speeds, close to the speed of light, they automatically pass into the future. The faster they travel, the further into time they move. However, it's not only muons that can travel in time. In principle anything can. The problem with sending a human being into the future is that an enormous amount of energy is required to accelerate an eighty-kilogram body to a speed close to the speed of light. Muons are very light and require little energy in comparison.

But imagine for the moment that we did have the technology to produce enough energy to accelerate a small rocket to a speed close to that of light. Further, imagine you have persuaded your mother to take a forty-year trip in this rocket. On the day she takes off you're eighteen, say, and she's fifty. Since she'll be up there for forty years, by the time she returns you will have been married and had a family. Your family will have had a family and you will be a grandparent aged fifty-eight. This of course would make your mother a great-grandmother aged ninety. Or would it?

On the day she arrives back you rush out to the airport to greet her. But to your surprise, instead of a woman aged ninety, your mother has hardly changed. Her hair is perhaps a little greyer and her waistline a little thicker but apart from that she looks the same as the day she left. In fact she's been keeping track of time on a calendar and insists she's only been away for a few years. According to you, your mother has been away for forty years. According to your mother it has only been five. Although it is difficult for our minds to accept, both of you are correct. The reason for the difference is, because of the high speed of her travel, your mother has moved forward in time. She is only fifty-five years old while you're fifty-eight. Her own child is now older than her! The only catch is, however, since relativity only allows travel forward in time, once sent into the future there is no way of returning.

The reason it's difficult for our minds to conceive is because these things aren't common sense. But all common sense means is common everyday experience and, since these things only happen at very high speeds, they are not everyday experiences.

Although some of these side-effects, like time travel, seem weird, today there is ample proof that they happen. Very accurate clocks placed in high-speed jets have shown Einstein's time travel to be true. In fact there are pieces of machinery that simply wouldn't work if relativity were not correct.

Not long after releasing his first bombshell, Einstein proposed another theory that was equally revolutionary — his General Theory of Relativity. The predictions from this theory are equally bizarre. Imagine you were capable of throwing a stone with incredible speed and strength. In fact you can throw a stone so fast that, like a rocket, it escapes the Earth's gravity, whizzes past the Sun and planets and heads out into uncharted space. The question is, assuming the stone goes on for ever, where does it end up?

General Relativity predicts that, in theory at least, it could end up hitting you in the back of the head! That the stone could travel right round the universe and end up back in the place it started.

Now it's an important point that the stone never actually turns around or even slowly curves in the normal sense. It always heads in a straight line away from Earth. The reason it actually ends up back at Earth is because Einstein's General Theory of Relativity predicts that the universe itself is curved, that the actual three-dimensional space we live in could be bent around, forming an enormous four-dimensional sphere. And just as the Earth can be circumnavigated by simply following a straight course, so can the universe.

It is impossible to visualise a curved three-dimensional space, or a 4D sphere for that matter, so don't try. The problem is that we are confined to three dimensions. To see the curvature would require looking down from a fourth dimension, impossible for us 3D creatures. The problem would be the same for a two-dimensional creature confined to a 2D world. It wouldn't be able to perceive the curvature of its world without looking down on it from a third dimension. For example, if its two-dimensional world was the surface of a ball, the only way it could see that it was curved would be to leave the surface and look down on the ball. Just as the only way we can see that the Earth is not flat is to view it from space.

The idea that the universe is curved is accepted by almost all astronomers. The question is, however, whether it is shaped like the surface of a 4D ball or more like the surface of a 4D saddle. It would only be possible, of course, to circumnavigate a closed surface like that of a ball. The answer depends on the total amount of matter in the universe. And so far only a fraction of the matter needed for the universe to be closed has been found. So at present the question of whether the universe is closed must remain open.

In practice, circumnavigation of the universe would actually be impossible. The reason is that nothing can travel faster than light and even something travelling at this speed would take longer than the age of the universe to complete the trip.

# $F$ire-walking

$W$alking barefoot over red-hot coals is an experience I can highly recommend. It is truly an amazing feeling strolling over a 3-metre bed of glowing embers and not feeling the slightest pain. I tried my first fire-walk in the backyard. No, it wasn't a barbecuing accident or even the result of a drunken bet. It was an attempt to really put science to the test.

The very idea of walking over red-hot coals seems crazy to most of us westerners. However, there are still many cultures around the world today that practise the art of fire-walking. In fact fire-walking is one of the latest crazes to sweep through California. Its proponents claim it is only a matter of training your mind to overcome the coals. In fact, for around a hundred dollars, a fire-walking guru will give you all the necessary training and escort you through your first fire-walk.

However, walking over red-hot coals is not as difficult as it might seem. In fact the ability to walk on coals can be explained by some very simple physics.

6

It was an attempt to put this physics to the test that had me lighting a raging bonfire in my backyard. Once the flames had died down I shovelled the red-hot embers on to the lawn, forming a neat 3-metre-long strip. Just for the record, I measured the temperature of the coals before I took my first step. The thermometer shot up to 310 degrees Celsius! Not particularly comforting, considering that water boils at 100°C and human flesh starts to burn at about forty-five degrees.

Taking that first step was hard. Even though I had faith in the physics, the glowing embers seemed very forbidding. My feet looked so tender next to those coals. I held my breath and took the plunge. The second, third and fourth steps followed very quickly and before I knew it I was standing on the grass at the other side of the coals. A feeling of elation overcame me. I had just walked across a bed of coals reliably measured at over 300 degrees without suffering so much as a blister!

Now before I go any further I should warn you that fire-walking can be dangerous. I don't recommend that you rip the hot plate off your barbecue tonight and start dancing around on the burning embers. I suffered a few burns before I got everything right. And there are a few things you do have to get right. For example, it is important to scoop the coals from the fire on to the lawn before walking on them. Temperatures inside the bonfire can be extremely high and if you attempted to walk on the fire itself you could be severely burnt. Another trick is to wet the grass surrounding the bed of coals. This lowers the temperature of your feet before-hand and helps to add a little extra protection.

However, once you have mastered the fire-walk, it is then incredibly easy. You can stroll across the coals at a leisurely pace without the slightest hint of pain. In fact, most of the time your feet don't even feel warm!

The scientific explanation is simple. Although the coals are certainly over 300 degrees, my feet never get anywhere near this temperature. In fact they don't even get to forty-five degrees. After all, if they did I would have burnt feet. The fact is that coals give off their heat very slowly. So I have time to walk across them before they have time to heat my feet past burning point. If I stood on the coals for ten minutes straight there is no doubt about it — I'd cook!

There is an important difference between heat and temperature. For example, take a metal frying-pan with a wooden handle that has been in the oven for a couple of hours. The metal and the wood must be at the same temperature, oven temperature. However, the metal will burn you much more readily than the wooden handle. This is because metal gives off its heat much faster than wood. It can, therefore, raise the tempera-ture of your hand to burning point very quickly.

Some scientists also believe that the wet grass helps with the fire-walk. As the water evaporates from the feet, the so-called Leidenfrost effect could add some insulation. The best way to observe the Leidenfrost effect is to tip a glass of water on to a barbecue plate. Instead of evaporating instantly, bubbles of water roll about on the plate for some time. They are insulated by a thin layer of steam that forms between the bubble and the hot plate. It is thought that the same effect might give the wet feet of fire-walkers some added insulation. As the water evaporates, steam may become trapped between the feet and the coals.

Although physics can explain why my feet don't burn, the mind does play some part in fire-walking. While I don't believe that mind control can stop physical damage like burns and blisters, the mind has complete control over pain and fear. In fact one night I did a fire-walk for a camera crew and my feet didn't even get warm. Intrigued by how simple it seemed, an inquisitive cameraman gave it a try. He walked across screaming! I remember the first time I tried it I also felt some pain. Your mind expects to feel pain and as soon as it feels anything it interprets it as pain. In fact, all that this cameraman was feeling was the crunching of charcoal under his feet. Once your mind understands this the pain mysteriously disappears.

# Wanted — a demon-proof Esky!

*I*f the chair you were about to sit in suddenly took evasive action and leapt to the other side of the room, you would think that either your house or your mind was haunted. But the disturbing possibility that one day the ham could suddenly leap off your sandwich or your normally docile settee could hurl itself through the television is actually predicted by physics! In fact physics makes some even more distressing predictions. Imagine submerging your warm tinnies in an Esky of crushed ice only to find that the beer gets warmer and the ice colder!

These predictions date back to early last century, to a time when physics was trying to come up with a theory for temperature and heat. One theory was that heat was some sort of fluid that supposedly flowed out of hot objects and into cold objects. However, this idea had problems. For a start, all fluids must be made of some material substance. What sort of material could heat be made from? But an even greater problem

was that heat didn't obey one of the basic laws for fluids, the law that matter can neither be created nor destroyed, except in nuclear reactions. Heat can be created and is created every time a fire is lit or a heater turned on.

A better theory for heat came from another revolutionary breakthrough that was happening around the same time: the discovery of atoms. The idea that all material objects are composed of basic indivisible building blocks, or atoms, had been around for thousands of years, since the ancient Greeks. However, it was only early last century that evidence for the existence of atoms started to appear. As well as providing a radically different understanding of the structure of matter, atoms also gave credence to a revolutionary new theory of heat. This theory stated that heat was nothing more than vibrating atoms. The more an object's atoms vibrated, the hotter it was. The atoms in a hot saucepan, for instance, shake vigorously, while those in a block of ice vibrate much more slowly.

The atomic theory gave a beautiful description of heat which stands to this day. However, it also revealed the possibility of a few weird side-effects. For a start, if every object is made of vibrating atoms why doesn't a coffee cup shake itself off the dinner table? The simple answer is that it can!

The vibrating atoms in every object are held together by forces. A good analogy might be to tie together a group of hyperactive school children, although I wouldn't try — it's probably illegal. Each of the children will run around vigorously, first pulling in one direction and then in another. However, despite the frantic movements of each child, the group overall will hardly move. The reason, of course, is that each time a child pulls in one direction, there is bound to be another pulling in the opposite direction. Overall, all this pulling and pushing tends to even out. The more children, the more chance of evening out.

Since in a coffee cup there are an enormous number of atoms, all the atomic vibrations are completely evened out and the cup doesn't move. Well, this is the case most of the time anyway. Since atoms vibrate randomly, there is a slim chance that a group of them could vibrate in the same direction at the same time. And if enough atoms were involved, the cup could lift off the table, coffee and all! But don't panic and start tying down all your material possessions. The chance of such a tremendous fluke is unimaginably small. You can never hope to see it.

Similarly, tricks can happen when cold objects come into contact with hot ones, for example, when warm beer comes into contact with ice. What normally happens is that the ice cools the beer and the beer warms the ice. If left long enough, the two eventually reach the same temperature. In terms of atoms, what happens is that the vigorously vibrating beer atoms collide with the not-so-vigorous ice atoms. The fast atoms slow down while the slow ones speed up. If left long enough, both the beer and ice atoms eventually vibrate at the same speed, corresponding to the two reaching the same temperature. However, there's a chance that something completely different could occur.

Since atoms vibrate randomly, the possibility exists that when the beer and ice come into contact, the collisions will cause the rapidly vibrating atoms to vibrate even more rapidly and the slower ones even more slowly. The beer would then get warmer and the ice colder. Instead of the ice cooling the beer, it would warm it up! However, once again the chance of this happening is so small you would never expect to see it. That is, unless one of Maxwell's demons was present.

Early last century a physicist, James Clerk Maxwell, had a devious thought. Suppose an intelligent creature the size of an atom invaded an Esky. By simply allowing some atoms to vibrate while stopping others, the miniature demon wouldn't have to wait for the small random chance. It could warm your beer whenever it desired. Fortunately these antisocial creatures remain purely fictional.

The vibrating atom theory of heat brought with it another remarkable prediction. Suffering in the winter cold, it's easy to forget how warm it gets in some parts of the world. But these temperatures, of course, are nothing compared to those inside the sun or in an exploding galaxy. It seems that there is no limit to how hot some parts of the universe can get. However, the atomic theory predicts that there is a very definite limit to how cold things can get.

The temperature on the moon at night is about $-150°C$. In the depths of outer space it gets down to about $-270°C$. But it turns out that nowhere can get colder than $-273°C$. This is the absolute minimum possible temperature. It occurs when the atoms stop vibrating altogether. Once the atoms stop, they can't slow down any further, so the temperature can't drop any lower.

# *Cure by mind power*

*E*vidence that the mind can influence the body's reaction to disease has been growing in recent years. In his latest book, *The Mind Machine*, Oxford University Professor Colin Blakemore relates a fascinating story about the power of the mind over cancer. The story is about a patient back in the 1950s who was suffering from terminal cancer. As treatment, the man was receiving a drug called Krebiozen. Remarkably, he responded well to this treatment and underwent an amazing remission. However, not much later, reports appeared saying Krebiozen was useless in the fight against cancer, that it had no therapeutic effect whatsoever. Suddenly the man relapsed into serious illness. In a desperate attempt to rectify the situation his doctor gave him distilled water, telling him it was a very pure form of Krebiozen, known to be effective. Once again the man made a remarkable recovery. However, two months later a particularly scathing report on Krebiozen was released. The man died within a few days.

Medical science is becoming increasingly aware that the attitude of the patient is a factor in the fight against disease. The link may be through the body's immune system.

The immune system is all that stands between us and a hoard of threatening microbes. A condition known as Severe Combined Immunodeficiency (SCID) is a sad example of the problems that can occur when the immune system is deficient. Most children with SCID die when they are less than a year old. The only way they can be kept alive for longer periods is to be confined to a sterile environment. Doctors in Texas kept one boy alive for twelve years by putting him inside a sterile glass bubble. The bubble simply served to keep out bacteria, viruses and fungi. All the boy's food was sterilised and the air filtered. The hope was that he could be eventually cured with a bone marrow transplant. Unfortunately the operation was unsuccessful and the boy died.

It had always been thought that the body's immune system was automatic and lay outside the control of the mind. However, in recent years there has been increasing evidence that the human mind can have some power over the immune system. In fact there is a whole new area of brain research, neuro-immunology, specifically concerned with this subject. The research indicates that the mind can actually suppress the immune system.

Some studies of widowers have revealed that parts of the immune system can be suppressed in the year following the death of their wives. Similar results have also been found for people caring for a close relative with Alzheimer's disease and for seriously depressed patients. In fact research indicates that even animals can learn to suppress their immune systems.

Under normal conditions, rats love the taste of sweet water and if they have the chance they will eagerly drink it. However, in an experiment designed to change this, a group of rats were fed sweet water, followed by a dose of a drug that suppressed their immune system and made them feel nauseous. Sure enough, after a while the rats learnt to avoid sweet water. However, some time later, some of the rats went back to drinking the water. And surprisingly, although no drug was administered, the rat's immune system still became suppressed. The rats were actually suppressing their own immune systems in response to the sweet taste! Through conditioning, they had learnt to associate the taste of sweet water with the immuno-suppressive drug.

The fact that a treatment with no therapeutic value at all can have a dramatic effect on a patient is known as the placebo effect. For the effect to take place, it is important that the patient believes the treatment will work. The relief of pain is a good example. In his book, Blakemore relates an experiment conducted by Jon Levine from the University of California. The experiment involved a group of patients who had just had their wisdom teeth removed. The group was divided in two and all of the patients were told they would receive an injection to reduce their pain. They were not told that the injections only contained a saline solution. For the first group, the injections were administered by a faceless automatic pump controlled by a computer. While for the second group, they were administered by Levine himself. Levine, wearing a white coat and carrying a stethoscope, carefully administered the injection, making sure he was in full view of each patient.

After the experiment each of the patients was asked to rate the pain level on a numerical scale. The group for which Levine administered the injection himself actually suffered much less pain than those injected by the machine. Levine believes the image of the doctor in a white coat with a stethoscope has an important placebo effect in our society. It is thought that the effect is caused by chemical substances in the brain similar to morphine. On the expectation that pain will be reduced, the brain produces morphine-like substances that do actually block the pain.

Blakemore goes on to say that he believes that miraculous cures, like those at Lourdes in France, are also caused by the placebo effect. Lourdes was made famous in 1858 by Bernadette Soubirous when she claimed to have seen a vision of the Virgin Mary. In the years since this apparition, the town has become known for its miraculous cures. People now visit in their millions. And apparently miracle cures have been occurring — officially sixty-four, according to a local medical bureau. Blakemore says that simply believing that Lourdes can cure does cure some people.

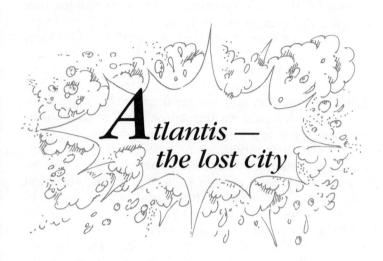

# *A*tlantis — the lost city

*A*s we go about our daily affairs, going to work, mowing the lawn, scrubbing the bath, it's easy to think that nothing much happens in life. It's difficult to entertain the possibility that one day the whole of our country and its civilisation could suddenly vanish in a single cataclysmic disaster. However, history tells us that major catastrophes like this do happen.

In fact, roughly 3,500 years ago, one of the most developed civilisations of that time did vanish. The catastrophe occurred on the Greek island of Santorini, about a hundred kilometres north of Crete. Its cause was an enormous volcanic eruption so large it blew out the centre of the island, completely destroying all signs of civilisation in the process. In fact much of the original, roughly circular, island disappeared during the eruption. Today all that remains is a crescent-shaped piece of land outlining part of the original isle.

The Santorini eruption was one of the great cataclysms of the ancient world. It was even more significant than the eruption of Mount Vesuvius that covered Pompeii in AD 79. In fact it is the most likely source of the legend of Atlantis, first told by Plato. Plato wrote that Atlantis had been a powerful and wealthy island whose people originally had a peaceful nature. However, this all changed when they decided to set out and conquer the Mediterranean. They were eventually defeated by the Athenians and as punishment for their aggressions, the legend goes, the gods destroyed their civilisation with earthquakes and floods and their city sank into the sea. Plato said he based his story on certain ancient accounts. It could have been that news of the drama was recorded in ancient Egypt. After all, the Egyptians knew Crete, or Keftia as they called it, quite well.

Today the volcano still smokes and steams, occasionally letting out larger belches. The last eruption occurred in 1928. In apparent defiance of the volcano, the present-day inhabitants of Santorini perch their buildings at the top of the 300-metre-high cliffs that formed during the island's destruction. The views offered by these cliff tops are nothing short of spectacular. Sipping coffee at one of the island's cliff-top cafés is an experience that shouldn't be missed!

Although the eruption was devastating for Santorini it may have triggered a more far-reaching disaster. Around the same time as the Santorini eruption, the great Minoan civilisation on nearby Crete also crumbled. In fact the most recent evidence dates the eruption to a time very close to the downfall of the Minoans.

There are many theories on how the Santorini eruption could have affected Crete. For example, earthquakes or tidal waves generated by the explosion might have wreaked havoc on the island. Even an air percussion wave following the eruption could cause buildings to collapse. Storage vessels would break, oil would spill and widespread fires could follow.

However, all of these theories hinge on one thing. It must be shown that the eruption and the demise of the Minoans occurred at the same time.

The scientific detective work that has linked these two events is fascinating. And although the eruption occurred 3,500 years ago, scientists are still uncovering new evidence of the effects it had on the world. Some of the evidence turns up in what you might think are the most unlikely places. For example, evidence for the eruption has been found in the growth rings of the exceptionally long-lived bristlecone pines of southern California. The connection is through the climate.

Large volcanic explosions can affect climates around the world. During a volcanic eruption, large amounts of dust and ash are blown high into the earth's atmosphere. At high altitudes, atmospheric circulation is small and therefore the air is fairly stable. As a result volcanic debris at this altitude can take years to settle back to the ground. While it remains in the atmosphere, it blocks some of the sun's radiation and causes lower temperatures on Earth.

Although this reduction in global temperatures would have been quite small, it was enough to affect the growth rings of trees around the globe. The growth rings are smaller during times when temperatures are lower.

Two American researchers, Valmore LaMarche and Katherine Hirschboeck, examined the bristlecone pines and found only one period in the middle of the second millennium BC when the growth rings were significantly affected. They calculated this to be 1627 BC and believe that it was caused by the Santorini eruption. An examination of the rings of ancient trees preserved in the Irish peat bogs also supports this conclusion. Low growth rates occurred in the 1620s BC.

Evidence of a different kind has been found in Greenland, not in the tree rings but in the ice. Just as tree rings record the years as layers of growth, the layers in the ice sheets of Greenland do a similar job. The reason is that each year's snowfall forms a new layer of ice on the existing ice sheet. If you drilled into the ice sheet you would find that the deeper you drilled, the older the ice would be. If you drilled down deep enough, you would eventually find ice that is 3,500 years old — ice that formed just after the Santorini explosion. A Danish scientist, Claus Hammer, and his colleagues have done just this. They examined the ice layers in Greenland and found an especially acidic layer of ice that was laid down within twenty years of 1645 BC. They believe it was caused by the Santorini eruption.

Further evidence dating the Santorini explosion comes from radiocarbon dating of twigs and grain and from archaeological evidence. Taking all the evidence together, it seems likely that the destruction of Santorini occurred at the same time as the disappearance of the Minoans. It is very tempting to say that one caused the other.

# *T*he day it rained frozen ducks

On 13 November 1976 an elderly Australian couple suffered an extraordinary death. They were sucked from their car by a passing tornado. Eyewitnesses reported that the tornado lifted the car nine to twelve metres into the air before dumping it in a ditch more than a hundred metres away. Both bodies were later found, stripped naked by the fierce winds, about sixty metres from their car. The passenger side seatbelt was still buckled.

We hear about the well-publicised tornadoes that tear through the Great Plains of the United States every year. However, it is not well known that tornadoes (not to be confused with cyclones) are quite common in Australia. In fact, next to America, Australia has more tornadoes than anywhere else in the world. The freakish deaths of November 1976 did not occur in America but in central Victoria!

Tornadoes occur in every state in Australia. They have a characteristic funnel shape that reaches down from the clouds to wreak havoc on the ground. And it is havoc indeed. The fierce winds in the core of a tornado have been known to lift a seventy-tonne railway carriage off its tracks and blow it more than twenty metres. And the bizarre stories don't stop there. There are tales of tornadoes stripping the feathers off chickens without killing them, and accounts of torrential downpours of fish and frogs. And many reports of exploding houses and storms with hailstones the size of grapefruit. There are even instances where blades of straw have been found embedded in wooden fenceposts! Tornadoes also kill. In 1925 a single tornado, or twister as they are called in the United States, passed through Missouri, Illinois and Indiana, and killed 689 people.

Tornadoes are very different from cyclones. Cyclones typically affect a region 800 kilometres across while the concentrated whirling winds of a tornado are only a few hundred metres across. Because of their size, jumping in a car and driving is quite a feasible way to avoid an approaching tornado. Usually tornadoes only last for about half an hour, carving out a path of destruction twenty-five kilometres long. However, there can be major exceptions to the rule. The 1925 tornado in the US tore a track 350 kilometres long and in some places up to 1.5 kilometres wide!

The death and colossal destruction caused by tornadoes year after year have meant that tornado research is a high priority. But even today, the details of just how and when tornadoes form remain enigmas. There is certainly no reliable way of predicting when and where they will strike. Part of the problem lies in the fact that it is difficult to take direct measurements of tornadoes. Meteorological instruments are torn apart by the whirling winds.

However, although the detailed workings of tornadoes are not known, some of the basics are understood. The willy-willy is similar to a tornado in a number of ways. Willy-willies, or dust-devils, are a common sight on sunny days. Even in the school ground I remember tight whirls of wind that would lift dust and Twisties packets high into the air. Two ingredients needed to create a willy-willy are sunshine and rotating winds. Rotating winds are created every time the wind blows past an object, although often the rotation is so slow you don't notice it.

The other vital requirement for a willy-willy is sunshine. Sunshine heats the ground and this warm ground heats the layer of air above it. Since darker colours get hotter than lighter ones, pockets of hot air form over the darker patches on the ground, for example, over a black bitumen road or a dried-out dam. Once hot enough, the warm pocket of air rises and that's when the willy-willy begins. It's the combination of rising air and rotation that can amplify a slowly revolving wind to a whirling willy-willy

capable of lifting dust and occasionally even small rabbits high into the air.

The reason can be seen by watching a spinning ice-skater. Picture a figure skater spinning on the spot with her arms outstretched. As she lowers her arms, eventually bringing them down to her sides, she spins faster. If she raises them again she slows down. It's the drawing in of her arms that makes her spin faster. In the same way a rising pocket of warm air draws surrounding air into the space behind it. Since the surrounding air is slowly rotating it speeds up as it is drawn in. Before long a rapidly whirling willy-willy has formed.

The same potent mixture of rotation and updraft in a willy-willy also operates in a tornado. Tornadoes almost always occur during thunderstorms. The familiar black towering clouds associated with thunderstorms are actually formed from condensation drawn up by enormous updrafts of warm air from the ground. It is these updrafts that give thunderclouds their fierce reputation among hang-glider pilots. If a hang-glider flies near a thundercloud it can be sucked in, turned into an enormous hailstone and rained back to earth. There are stories of frozen ducks in among large showers of hail.

As warm air rapidly rises during a thunderstorm it is set into rotation by the upper-level winds and if the conditions are right a fierce tornado is spawned. Often several tornadoes appear during the one storm. The rapidly whirling winds mean at the centre of a tornado the pressure is very low. This low pressure can literally cause houses to explode as the tornado passes over, especially if the windows and doors are closed and the house is well sealed. The situation is a bit like a pressure pack can on an aircraft. Because pressures are low high in the atmosphere, pressure pack cans in the baggage department can explode.

As for the torrential downpours of fish and frogs, they are caused by tornadoes passing over water. Fish and frogs have been known to be sucked up and rained back over dry land.

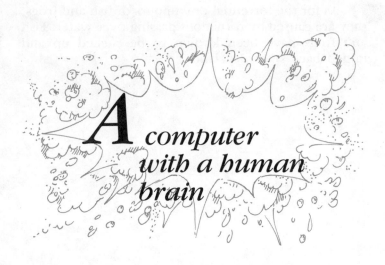

# A computer with a human brain

*T*he quest to develop a computer more intelligent than man has been going for some time now. However, in recent years a radical new approach is being investigated. Rather than just simply trying to build bigger and faster machines, computer scientists are now experimenting with a new breed of computer — a computer actually designed along the lines of the human brain.

In many ways computers can already outperform man. Even a five-dollar pocket calculator is embarrassingly superior to the human brain when it comes to multiplication or division. However, there are still some tasks for which the brain wins hands down, for example, the ability to recognise a face from a photograph or the species of a particular tree in the park. These are simple tasks for the human mind, but extremely difficult for even the most advanced computers.

The reason is that computers are not good at recognising patterns. Recognising trees, faces or any complicated object for that matter, is essentially a problem in pattern recognition. Computers are only well suited to tasks that have a clearly defined set of rules. Good examples are division and multiplication. Both of these are governed by the simple rules learnt by every child in primary school. In the same way as a child, a computer multiplies and divides by simply applying these rules. The rules are the same whether multiplying two three-digit numbers or two fifteen-digit numbers.

When it comes to recognising trees, however, there are no precise rules. It is impossible to write an exact definition that covers all possible trees. If you don't believe me, try it. But you must remember, a computer is just like an alien being that has never seen a tree before. You will have to start right at the basics, telling it absolutely everything. It has never heard of branches, leaves or even bark.

It turns out that the only way to get a computer to recognise any conceivable tree you might show it

is to program every possible tree into its memory. The computer then simply checks through everything in its memory until it finds a match with what it sees in the picture. Although this all sounds quite feasible, there is a problem. The search could take an enormous amount of time. After all, it could be that the very last tree in the computer's memory is the match. The human brain, in comparison, only takes a fraction of a second to recognise patterns. It is this ability that has prompted computer scientists to turn to the brain for new ideas.

The human brain is an extremely sophisticated computer. However, in many ways it is very different from a conventional computer. For example, it stores memories differently. A computer stores memories on a magnetic disc, each individual memory being stored on a different part of the disc. If you took a knife and scratched the section of disc that contained a particular memory, that memory would be destroyed. However, human memory is quite different. A surgeon can actually remove parts of the brain without affecting the memory. The brain must, therefore, store memories in a very different way. Efforts to understand how the human memory works have led to other possibilities for computer memories.

For example, it is possible to store memories in the form of patterns. A large multistorey building is a simple example. Imagine you had a remote-control panel that could operate the lights in each room of the building separately. By simply switching the lights on and off in the various rooms, many different patterns could be formed on the outside of the building. These lights could be used to store memories. For instance, I could store my age, which unfortunately has just become thirty. I would simply switch on the light in the first room on the thirtieth floor and leave all other lights turned off. Similarly, an age of forty could be stored by turning on the light in the first room on the fortieth floor. Providing there were enough floors, every possible age could be stored in this way.

This is very similar to the way a conventional computer stores memories. Information is stored in specific locations, in this case in specific rooms. However, there is another way to store these ages. Instead of representing a memory by a single light, it is possible to store each age as a pattern of lights. For example, you could switch lights on and off on many floors, forming a triangle on the front of the building, which could represent the number thirty. The number forty could be represented by a circle. In fact a different pattern would be used to store every possible age.

Both of these techniques can store all possible ages, but the use of patterns has a very strong advantage. When patterns are used, each memory is spread over the many lights in the pattern. This means that a few faulty light bulbs don't affect the memory. If one light bulb blows, the circle is still recognisable as a circle and the triangle as a triangle. However, the first technique, used by conventional computers, is very susceptible to faulty light bulbs. If the light bulb in the first room on the thirtieth blows, the memory goes with it.

The human memory seems to be like the second type. It uses patterns to store memories. The brain consists of an enormous number of interconnected neurons, each of which is connected to many others via electrical and chemical signals. Rather than being stored in individual neurons, the memories reside in patterns of neural connections. The number of patterns that can be formed from this tremendous number of neural connections is mind-blowing. It makes you wonder just how the brain can sift through its enormous collection of memories and recognise a face in a picture so rapidly. Fortunately, this is another area where the human brain is superior to a computer. It doesn't have to search through every possibility.

Instead the brain has an associative memory. Once it is given some information it uses that information to recover the complete memory.

Computer scientists are starting to experiment with computers that have associative memories and memories that use patterns of connections for storage. Although these neural computers are still very crude compared to the human brain, they have already had some success with pattern recognition problems. However, work in this area has only just begun. The real advances are yet to come.

# *B*leak weather ahead

*E*veryone knows there have been times when the Earth has suffered great ice ages, times when snow and ice have covered large parts of the globe. However, it is not so well known that our planet is currently poised on the edge of another ice age. For the Earth to experience a full ice age a number of conditions must be present. Currently, most of them are. A long run of severe winters might be all that is needed to topple the scales and trigger the onset of a new ice age.

A run of such winters could happen at any time. It may happen next century. It could take a thousand years. But one thing is certain: it is bound to happen some time in the next 4,000 years. In fact ice age conditions have been upon us for some time. We should think ourselves lucky that a run of savage winters in the past hasn't already triggered a new ice age.

Ice ages have been coming and going for millions of years. In comparison, civilised man has been on the

Earth for a mere 10,000, a blink of an eyelid in comparison. To put these times in perspective, imagine the whole of the Earth's history squashed into one hour.

At the beginning of the hour the Earth was born and within fifteen minutes early life was well under way. After this rapid beginning things then appeared to slow down a little. For the next half an hour or so the highest form of life on Earth was seaweed. It was only during the last eight minutes that more complicated life developed. During these eight minutes the first real animals evolved. They developed first in the oceans and then made their way on to the land. During the last three minutes the dinosaurs appeared. They ruled the world for two minutes and then vanished into extinction. It was only in the last couple of seconds that man appeared. And only in the last one-hundredth of a second that he moved out of caves and farmed the land. This was about 10,000 years ago. The last ice age ended at about this time. In fact the end of the ice age and the subsequent rise of civilisation was probably no coincidence. The warmer weather may have had a lot to do with mankind's rapid development from nomad to urban dweller.

This most recent ice age was by no means the first. In fact ice ages have flickered on and off regularly during the last few million years. In compressed time, each ice age lasts about a tenth of a second and the warm spells for about a hundredth. So for the last few million years, normal weather for planet Earth has been a full ice age. The warm spells between have been very brief. We are lucky to live in one!

The Earth, however, has not always been plagued by regular ice ages. Moving two minutes back in time to when the dinosaurs ruled, average world temperatures were much warmer than today. Ice ages were non-existent, the north and south poles were completely free of ice and lush tropical jungles were abundant. In fact it was this profusion of tropical growth that eventually formed the coal and oil deposits we rely on so heavily today. Strangely, the reason there were no ice ages then was due to the position of the continents.

Every continent and island on the Earth is on the move. Australia is slowly drifting northwards towards Asia. South America and Africa are drifting further and further apart. And Africa is closing in on Europe. As a result of their continual movement, from time to time continents drift over the north and south poles. This is what causes the Earth to enter an ice epoch. And it is during an ice epoch that we suffer periodic ice ages.

When dinosaurs reigned there was no land near the poles and therefore no ice ages. However, today we live in a double ice epoch, a time when ice caps are at both the north and south poles. While there is only land over the southern pole, the crowd of continents surrounding the Arctic Ocean at the north pole has the same effect. They block the circulation of warm water and, therefore, lower the temperature.

When there is no land near the poles warm currents from the middle latitudes circulate through the polar regions. Any snow that falls melts as it lands in the ocean. However, when there is a continent over the poles, snow accumulates on the land. Since the warm currents can't melt this snow it builds up each winter. These permanent polar caps eventually lower temperatures around the globe.

The actual cause of individual ice ages within an ice epoch lies 150 million kilometres away in space. The most convincing theory for the cause of ice ages is the relative positions of the Earth and Sun. Once a year the Earth completes its orbit around the Sun. However, rather than the north and south poles pointing straight up and down, the Earth is tilted twenty-three and a half degrees. This, of course, is the reason for the seasons. At times of the year the northern hemisphere is tilted towards the Sun while six months later the southern hemisphere is. However, over tens of thousands of years the Earth's tilt varies and, as a result, at certain times the seasons are more extreme. In addition to this the shape of the Earth's orbit also varies. At times the Earth is abnormally close to the Sun, causing more extreme summers. It is extreme summers caused by a combination of these effects that melt the ice sheets and break the ice age. Remember, we are not looking for a reason for ice ages but for a reason why they melt. During an ice epoch a full ice age is normal weather.

Extreme summers have caused warm weather over the last 10,000 years. But the good times are almost over. The current tilt and the position of the Earth's orbit is that of an ice age. Once the ice sheets form the summers will not be extreme enough to melt them.

These days, however, there is a new player on the scene — the greenhouse effect. If greenhouse warming causes average world temperatures to rise significantly, ice ages of the future might be very different. There is even the possibility that they may no longer occur.

# The next ice age

$F$or the last 10,000 years or so planet Earth has been experiencing unusually warm weather. The summers have been warmer and the winters milder. However, it seems that the best of the warm weather is now over. We are heading towards a new ice age, a time when large parts of the land surface will be covered in ice, temperatures around the world will be lower and sea-levels will fall.

Ice age conditions have been normal weather for the Earth over the last few million years. Warm spells like the one we have been used to for the last 10,000 years are exceptions rather than the norm. And since our last ice age finished about 10,000 years ago, in geological time, the next one is due any moment.

Like almost all questions concerning the climate, making exact predictions is difficult. The Earth's climate is extremely complicated and comparatively little is known about the way its various parts interact. After all, making daily weather predictions is hard enough! However, judging from the past, the next ice age will almost certainly occur some time in the next few thousand years. And some idea of what it might be like can be gleaned by taking a look at past ice ages.

The peak of the last ice age occurred about 18,000 years ago. At that time about a third of the Earth's land surface was covered in ice. In comparison, only a mere tenth is covered today. If the conditions during the last ice age were to return, many of the world's northernmost cities would become covered in ice. Almost all traces of civilisation in Scandinavia would be wiped out. Helsinki, Copenhagen and Stockholm would simply disappear. Canada would suffer similar problems. It would be almost entirely overrun by ice and even cities as far south as New York in the United States would be severely affected.

In Europe, cities like Leningrad, Warsaw and Berlin would vanish under the advancing ice sheets. In the United Kingdom, Scotland, Northern Ireland and

even the north of England would become buried under hundreds of metres of ice. Even conditions in the ice-free parts of Great Britain would be inhospitable. The climate would be more like that in central Alaska today.

In Australia, the southern and mountain regions would succumb to the ice sheets. Large parts of Tasmania and southern New Zealand would be buried. The alpine regions of New South Wales and Victoria would also have an increased ice cover.

Cold weather and ice, however, are not the only changes that occur during an ice age. At the peak of the last ice age, about three-quarters of the Earth's fresh water was locked up in ice. As a result the average sea-level was about a hundred metres lower than it is today. Coastlines moved anything from one to 400 kilometres further away, increasing the land surface at the coasts. This was some compensation for the land that was lost to the ice sheets.

Even today a tremendous amount of water is trapped in ice sheets. Sixty million years ago, when there were no ice caps at the north and south poles, the Earth was relatively ice-free. Sea-levels then were about eighty metres higher than at present. A casual glance at an atlas shows that many of the world's cities would be wiped out if the polar caps were to melt and these sea-levels returned.

While looking to the past does provide some insight into possible conditions during a future ice age, there is reason to believe that the next ice age may be different. Human technology is a new variable in climatology. The human race is rapidly approaching a stage where we can make large-scale changes to the world climate.

The greenhouse effect is one glaring example. Man's activities have inadvertently added large amounts of carbon dioxide to the atmosphere which, if left unchecked, could result in a global warming of the entire planet. Greenhouse warming may play a vital role in the next ice age.

Another example of mankind's newly developed ability to make large-scale changes to the world climate received a lot of attention a few years ago. Atmospheric scientists discovered that even if only a fraction of the world's stockpile of nuclear weapons were used during a nuclear war, the climatic effects could be drastic. Updrafts created by the nuclear explosions would carry soot and debris high into the Earth's atmosphere. The effect would be partially to block the Sun's radiation, leading to cooler temperatures at ground level. Since dust and debris high in the atmosphere can take years to settle back to the ground, a worldwide winter lasting for more than a year could ensue.

However, mankind's ingenuity also has a more positive side. By the time the next ice age sets in, our technology might be sufficiently advanced to cope with, or even control, the ice sheets. Already scientists are examining ways of harnessing ice. Glaciologists are looking at efficient ways of taming the melt-waters from glaciers to fill reservoirs and for hydroelectric power. A great deal of thought has even been given to transporting icebergs from the southern ocean to the arid coasts of Australia and the Americas.

And although a new ice age would obviously mean some very drastic changes for some parts, these changes will be nothing that the world hasn't seen before. In previous ice ages many species simply migrated to the warmer climates of the equatorial zones. Others adapted to their new icy climate. The Eskimos are living proof of the human species' ability to adapt to an ice-covered world.

# *A*rmageddon

*T*he Sun will grow bigger, slowly changing from pleasant yellow to glowing red. Temperatures around the world will soar, parching the land and heating the sea. Gradually the oceans will boil and all life will perish. In the end the surface of the Sun will almost reach the Earth's orbit and our planet will be reduced to a charred wasteland.

Bleak predictions of the Armageddon are usually the domain of ancient mystics or strange religious sects. However, this one is a prediction of science. And unlike most predictions concerning the end of the world, the scientific one is actually supported by a great deal of evidence. There is no way of escaping the end. In fact the only consoling point is that the final days are not due for billions of years.

The Earth has suffered many great cataclysms in its time — collisions with asteroids, periodic ice ages and tremendous earthquakes. Life has managed to survive them all. Now we are creating our own cataclysms and nature is faced with new challenges. Environmental pollution is taking its toll, the human population is growing uncontrollably and global nuclear war is a real threat. However, there is no doubt that many forms of life will also survive these — even if human beings aren't amongst them! However, in the end, when the Sun expands and barbecues the Earth, there will be no survivors, no matter how tenacious the life form.

The Sun is just another star, one of the billions in our galaxy. Stars, like us, are born, they live and then die. They can be seen in all phases of their lives throughout the universe. At present, the Sun is roughly halfway through its life and enjoying a relatively peaceful middle age. However, once this midlife is over things become more interesting. The Sun enters its red giant phase. During this period its surface will expand past the orbits of the inner planets, engulfing Mercury and Venus. And although the Earth itself may not be engulfed, the Sun's surface will be so close that temperatures on our planet will be horrendous.

Burning rapidly as a red giant, the Sun will eventually exhaust its nuclear fuel and the red giant phase will end. When a star's fuel runs dry it starts to cool. And as it cools, the pressure drops and the star can no longer resist the force of its own gravity. The Sun will literally collapse under its own weight. The collapse will continue until eventually a cold and very dense dead star remains.

There are three possibilities for a dead star. Depending on its size, it can end up as either a white dwarf, a black hole or a neutron star. Stars about the size of the Sun eventually end up as white dwarfs. A white dwarf is a very dense, relatively small star. In fact these stars are so dense that one matchboxful of white dwarf would weigh several tonnes. However, white dwarfs are relatively tame creatures compared to the other two fates for a dying star.

Stars much bigger than the Sun undergo a violent explosion during their collapse. Such an explosion, known as a supernova, is one of the most violent events in the universe. During a supernova explosion a star can shine millions of times brighter than usual and debris from the explosion can be ejected into space at thousands of kilometres per second. During the early hours of 24 February 1987 the world witnessed a particularly special supernova event. It was special because it occurred relatively close to us. In fact it was close enough to be seen with the naked eye. Supernova 1987A, as it has become known, occurred in a nearby galaxy, only a mere 170,000 light-years away. The last nearby supernova happened in 1604, during the time of Johannes Kepler, one of the fathers of modern astronomy. Unfortunately the telescope hadn't been invented at that time.

After the explosion, the dying star is usually still too big to settle down as a white dwarf. Its force of gravity compresses it to even greater densities. If the star is not too big, the force of the neutrons in the star's atoms can eventually halt the collapse. The star then eventually comes to rest as an incredibly dense neutron star. Typically, neutron stars are only about twenty kilometres across and one matchboxful weighs more than a mountain!

However, if the star is much bigger, it seems that no force in the universe can halt the crushing force of gravity. Large stars seem destined to become increasingly dense. Eventually the force of gravity becomes so strong that the star squashes its own matter out of existence. All that's left behind is a black hole.

The gravity inside a black hole is so strong that even light cannot escape. And since nothing can travel faster than light, this means that nothing can escape. A spacecraft that accidentally passed too close to a large black hole could pass the point of no escape. And once this point is crossed, the spacecraft and its occupants are destined to be sucked towards the centre of the hole and squashed out of existence.

In comparison to its death, the birth of a star is a relatively peaceful event. Stars are born inside giant clouds of dust and gas. These clouds, which are found throughout the galaxy, collapse under their own weight, slowly compressing the gas and dust in the process. As the collapse proceeds, both the pressure and the temperature rise. Eventually temperatures near the centre are so high that nuclear reactions start to occur. At this point a new star is born.

# Animals that talk

*T*hey say that there is nothing wrong with talking to the animals, provided that the animals don't talk back. However, there seems to be an exception for scientists. During the last couple of decades or so, a number of researchers have claimed that their animals do indeed talk to them. In an attempt to bridge the communication gap between animals and humans a number of scientists have been trying to hold a two-way conversation with some of the more intelligent animals. They have been endeavouring to teach languages that both man and animal can understand to chimps, gorillas and even dolphins. The idea is that once a common language is in place, we can start to find out what goes on in an animal mind.

Since animals naturally produce very different sounds from ours, trying to teach them English would be ridiculous. Instead, a number of alternatives have been tried. One of these is American Sign Language which is used by the deaf. It has been taught to both gorillas and chimpanzees.

At America's Stanford University, Penny Patterson taught sign language to a female gorilla named Koko. Koko mastered a working vocabulary of about 375 signs. She could tell Penny what she felt like eating and what games she wanted to play. And she also exhibited some amazingly human-like traits. Koko would lie, joke, even argue and trade insults. She also seemed to have a touching empathy towards fellow animals. Seeing a horse with a bit in its mouth she signed 'horse sad'. When asked why the horse was sad, she signed 'teeth'.

Another famous talking ape is Sarah the chimpanzee. Rather than using sign language, Sarah communicates with plastic symbols. And her teacher, David Premack, believes that her mastery of this symbolic language has given us remarkable insight into the mind of a chimpanzee. Sarah is particularly good at recognising relationships. For example, when shown half an apple and asked to match it, she is capable of selecting a half-filled jug of water as the match rather than a whole apple. When asked the question: 'Tin is to can-opener as lock is to what?', Sarah's selection will be the key.

Chimpanzees and gorillas are mankind's closest living relatives. We can see ourselves both in their looks and in their behaviour. For this reason perhaps, trying to communicate with them is easier than with other animals. However, this hasn't deterred scientists from trying other possibilities. A number of attempts have been made to open the communication channels with those intelligent mammals of the sea, the dolphins.

Studies have been carried out both in laboratories and with wild dolphins at sea. In fact one of the best places for studying the behaviour of wild dolphins is here in Australia. At Monkey Mia, in Western Australia, scientists from around the world study the dolphins that regularly venture into knee-deep water to receive fish from tourists.

Both sign language and a language of sounds similar to those used by the dolphin naturally have been tried. As with teaching any animal language, the idea is to teach the dolphin a number of 'words'. These words, whether squeaks or signs, can then be put together to form sentences. An animal can be said to be learning language if it can put together its own sentences.

However, just as there are ardent believers that man and animal are now talking, there are also just as adamant disbelievers. In fact many scientists say that the sentences constructed by apes like Sarah and Koko are no greater an achievement than the tricks taught to circus animals. They say that a combination of wishful thinking with sloppy science and good old monkey imitation have misled researchers into believing that

their animals can talk. They don't believe that Sarah and Koko actually understand what they are saying. They have simply learnt to put together a string of symbols.

Scientists have to be very careful when doing research in this area. This is clearly seen in the famous story of Clever Hans, the horse. Towards the end of last century, a Prussian aristocrat by the name of von Osten believed he had made a remarkable discovery. He believed he had discovered a thinking horse. Osten would simply hold up series of numbered cards for Clever Hans to read, and the horse would tap out the number shown on each card with his hoof. Osten had himself and many experts fooled for some time. However, one investigator had a bright idea. He decided to see how well the horse would do if Osten didn't know the answers. He got Osten to hold up the cards to the horse without first looking at them. Suddenly Clever Hans's cleverness disappeared. Somehow Osten had been unconsciously passing on the answers to his horse.

It turned out that horses are very good at reading body language. Clever Hans had learnt to paw the ground when Osten leant forward slightly to get a better view of his hoof. Similarly, in anticipation of the correct answer, Osten tended unconsciously to straighten himself. Rather than Clever Hans reading and understanding numbers, he simply learnt when to start and stop tapping.

I guess only time will tell whether the capacity to learn a language is a purely human capability. It would be nice to think that one day we could ask a gorilla or a dolphin what their lives are like. However, wishful thinking should never come in the way of science. The research must go on.

# *A*t the edge of existence

*L*ike a pair of magic glasses, there is an instrument that allows you to see billions of years back in time, an instrument that actually lets you observe events as they happen, even though they happened billions of years ago. The instrument is the telescope.

This is one of the quirks of astronomy. Because distances are large, even light, which travels at the fastest possible speed, can take many years to cross the vastness of space. Quasars, for example, are one of the most distant objects in the universe. Light from quasars takes billions of years to reach Earth. As a result, we see quasars not as they are today but as they were billions of years ago. In fact to see what they look like right now, we have to wait another few billion years.

Of course, even without a telescope, when you look up at the night sky you are looking back in time. Light from our closest neighbouring star, Alpha Centauri, takes four years to reach us. Even our view of the Sun is eight minutes old. Everything in the sky is a delayed picture from the past. However, it is quasars that offer the most distant trip back in time.

Quasars remain one of astronomy's great unsolved mysteries. They appeared on photographs taken through telescopes for years before they were noticed. They seemed nothing more than relatively nearby faint stars. However, the mystery began when a radio telescope was pointed in their direction. Quasars were producing inexplicably large amounts of radio waves. And even the radio waves themselves were peculiar. The examination of a star's radio wave spectrum gives some information about the various chemical elements present on its surface. When quasars' radio spectra were examined, the waves corresponded to no known element in the universe!

At first astronomers were puzzled. Surely quasars don't harbour exotic new elements that exist nowhere else in the universe? But what else could explain the unusual radio spectrum? The puzzle was finally solved by Dutch astronomer Maarten Schmidt. He realised that the unknown radio spectrum for quasars was actually the simple pattern for hydrogen, except that the entire pattern was shifted to a different frequency. This shift in frequency of a star's radio spectrum is common and is caused by the expansion of the universe. However, the unusual thing for quasars was that the frequency shift was enormous. It meant that quasars were out near the edges of the universe, billions of light-years away. This created a new puzzle.

The question now was how could such distant objects be seen from Earth. After all, measurements indicated that quasars were relatively small objects, no bigger than the solar system. For an object that size to be visible over such large distances it would have to be tremendously bright. An unimaginably colossal source of energy would be required to power such a bright object.

This is where the quasar mystery stands today. Theories that try to explain such an incredible source of energy are many and varied. One of the more widely accepted ones is that quasars can be explained by matter pouring into enormous black holes. A 'black hole' is a region of space where gravity is so strong that matter is completely crushed out of existence. Black holes sometimes form during the final stages of a dying star. Once a star has exhausted its nuclear fuel, it can no longer provide the thermal pressure to resist its own gravity. The star literally starts to collapse under its own weight. If it is big enough, its gravitational force becomes so strong that no force in the universe can stop the collapse. The star continues to shrink, packing its mass into an ever-decreasing space. Eventually this compacting crushes the star out of existence, leaving behind nothing but a tremendously strong gravitational field and a hole in the continuum of space and time. This hole is called a black hole because nothing can escape its clutches. Not even light.

Everything that passes close to a black hole feels its strong gravitational pull. Whether it be an unsuspecting spacecraft or interstellar dust and gas, the black hole's unsuspecting prey is slowly pulled past the point of no return and finally ripped apart and crushed from existence. As the black hole's overpowering gravity tears the matter apart, there is a tremendous release of energetic radiation, so brilliant that a black hole devouring gas and dust could be an explanation for quasars.

Another theory is that quasars are galaxies filled with exploding stars. In a normal galaxy, like our Milky Way, an exploding star is a relatively rare event. In fact the last one occurred in the Milky Way in 1604. However, the situation would be very different in a galaxy where the stars are packed very closely together. A single exploding star, or supernova as they are known, could trigger neighbouring stars to explode and a chain reaction of stellar explosions might eventually engulf the entire galaxy. The enormous energy output that would occur might be an explanation for quasars.

Some of the other suggested explanations for quasars are even more bizarre. Some postulate that they are regions of the universe where matter and antimatter are in mutual annihilation. Others suggest they may be white holes, the opposite of black holes. However, a quasar's energy output may be no mystery at all. It could be that there is another explanation for the frequency shift of the radio spectrum, an explanation that doesn't require quasars to be at great distances. If they are not out near the edges of the universe, quasars require only a moderate energy source to explain their brightness.

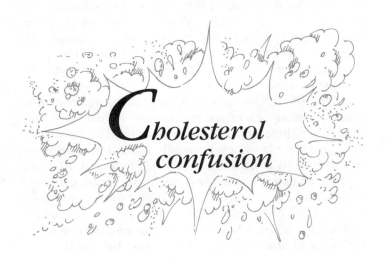

# *C*holesterol confusion

*C*holesterol would have to be one of the most talked-about chemical substances around. It seems that hardly a dinner party goes by without someone broaching the subject. However, despite this, there still seems to be widespread confusion about cholesterol and its connection with heart disease.

More often than not the subject comes up at dessert time, when a choice has to be made between the chocolate cake and the fruit salad. Reaching for two large pieces of cake, some argue that science hasn't proved cholesterol is a problem so why worry about it? Others adopt the Pritikin approach and steer clear of fats and oils completely. However, the confusion often goes even deeper than this.

A simple example is my favourite nut, the macadamia. In the dietary guidelines section of the Heart Foundation's booklet, 'Planning Fat-controlled Meals', they recommend that for a healthy heart, macadamia nuts should be avoided. Yet if you turn to the tables in the back of the book it states clearly that macadamia nuts contain absolutely no cholesterol. Similarly, while dieticians recommend a change to a diet lower in fat they also say that you should increase your intake of certain fats. It's not hard to see how the confusion arises.

First of all, contrary to what many people want to believe, most scientists no longer debate the role played by cholesterol in heart disease. The consensus view is that excess cholesterol is a killer. The more cholesterol in your blood, the higher the risk that you will die of heart attack. And one of the main causes of high blood cholesterol is diet. High blood cholesterol results in a build-up of fatty deposits on the artery walls. This white waxy build-up causes a narrowing of the arteries and heart disease can be the result.

However, to reduce your blood cholesterol level, simply reducing the amount of cholesterol in your diet is not enough. The reason is that there is a difference between dietary cholesterol and blood cholesterol. It's not necessary to eat cholesterol (dietary cholesterol) to raise the cholesterol level in your blood. Most people can raise their blood cholesterol by eating a diet high in saturated fat. This is because the body makes its own cholesterol — hardly surprising considering how important cholesterol is to life. It aids a number of body functions and is an important part of every cell wall. The problem, of course, is that although some cholesterol is vital, too much is dangerous. A diet high in saturated fat can lead to a dangerously high blood cholesterol level. This is the reason for the Heart Foundation's recommendation on macadamia nuts. Although they contain no cholesterol, they are high in saturated fat.

It seems that the generally high fat intake in the western diet is a major cause of the high heart disease rates in most western nations. In the heart disease stakes, Australia is roughly one-third of the way down the list, just above the United States and below New Zealand. Topping the list are Northern Ireland and Scotland while Japan and the southern Mediterranean countries are at the bottom. It is thought that the low heart disease rates in countries like Greece, Spain, Portugal and Japan are due to low blood cholesterol levels. However, fortunately, most people can reduce their blood cholesterol through a change in diet. The recommendation is a reduction in fat and an increase in cereals and grains. In dietary terms fat includes both solid fats and liquid oils. These fats come in three main varieties — saturated, monosaturated and polyunsaturated — the differences lying in their chemical structures. Of these three it is the saturated fats that increase blood cholesterol. For this reason dieticians recommend that, for a healthy heart, saturated fats should be replaced by unsaturated fats. The problem, however, is that most foods contain combinations of all three of these fats.

Animal tissue contains high levels of saturated fat and some animal products, especially eggs and shellfish, are rich in cholesterol. There are also vegetable products that contain large amounts of saturated fat. Cocoa butter, used to make chocolate, and coconut milk are examples.

Not everyone, however, can reduce the level of cholesterol in their blood sufficiently by diet. Some people are predisposed to high cholesterol levels. For these people, who have genetic disorders, drugs are the only solution.

The effect of diet on the body is an extremely complicated area. And consequently, finding 'absolute proof' that a diet high in cholesterol and saturated fats causes heart disease is not as simple as you might think. Instead evidence must slowly be gathered from many studies and experiments. For example, much information is obtained from national diets of different countries. The Japanese, for instance, traditionally eat a diet very low in fat. They also have an extremely low instance of coronary heart disease, which lends support to the idea that a diet high in fat causes heart disease. Further support comes from the Japanese that have emigrated to America. They soon acquire the American heart disease rate which is thought to be due to a change to a more western diet.

Because the whole question of cholesterol and heart disease is so complicated there are many issues that remain cloudy. Hopefully further studies will clear some of this fog. But in the meantime, follow the simple message from the Heart Foundation: cut down on saturated fats and cholesterol, and you will reduce your risk of heart disease.

# A new look at the missing link

*T*he search for the missing link between man and ape has always evoked controversy. In fact bitter argument is probably a more accurate description. However, recent advances in molecular biology have shed new light on this age-old debate. These advances have shown that, in terms of genetics, the chimpanzee and human being are very close. In fact, genetically, man and chimp have more in common than chimp and gorilla! When the genetic knowledge is added to the fossil evidence it now seems likely that the common ancestor between man and ape was a creature very similar to the chimpanzee.

Traditionally, the family tree of evolution has been pieced together from fossils. Every bone, tooth and skull ever found is used to form a grand picture of how life evolved on earth. The age of the bones determines just where each prehistoric creature fits on the family tree. The problem is that the fossil record is far from complete. Many more fossils would be

needed before a comprehensive picture of evolution
could be made. Another problem is that looks can be
deceiving. In a court of law, blond hair and blue eyes
are not enough to establish paternity. Blood tests are
required. The recent advances in molecular biology
have served as the equivalent of a blood test for evolution. Rather than blood, molecules are compared.

Each cell in an animal contains all the genetic
information required to make the complete animal.
This information is encoded in a molecule and carried
in the genes. The molecule is the molecule of life,
DNA. The DNA in each creature that ever lived is
unique. It is like a genetic fingerprint. In fact DNA
prints are even more powerful than fingerprints. They
can be used to give unequivocal proof of blood relatives in immigration disputes and can establish
maternity and paternity with 100 per cent accuracy.

When the DNA of man and ape are compared
interesting results are found. Traditionally, there have
been three candidates for man's closest living relative:
the chimpanzee, the gorilla and the orang-utan. From
the fossil evidence it is difficult to put one closer than
another. However, when the DNA from each species
is compared a pattern emerges. The orang-utan is the
most distant relative of the three. It split away from
our common ancestors almost twenty million years
ago. Then followed the gorilla approximately ten million years ago. Finally chimpanzees and man went
their separate ways between five and ten million years
ago. This makes the chimpanzee our closest living relative. It is a prime example of how looks can be deceiving. Based on appearances alone, you might expect
chimpanzees and gorillas to be much closer to each
other than to man.

Although genetic engineering has made radical in-
roads into anthropology and evolution it hasn't
replaced the study of fossils. Dating the fossilised
remains of prehistoric animals will continue to be a

prime instrument in the anthropologists' tool kit. The method used to date bones and teeth is known as radioactive dating and can be used to date anything from a bottle of vintage wine to the Earth itself. Recently carbon dating was used to determine the age of the Shroud of Turin, regarded by some as the burial cloth of Christ. Small samples of the shroud were given to three independent laboratories for radioactive dating. Unfortunately, rather than being 2,000 years old, each of the tests revealed that it could be no older than 726 years. As a test of the methods used, dummy samples of known age were also tested. One of these was the mummy cloth of Cleopatra and was dated between 9 BC and AD 78. It was good that each of the laboratories agreed so closely. On a previous test one of the laboratories was out by 1,000 years!

The process of carbon dating is simple. Carbon atoms come in two forms, referred to as C12 and C14. The vast majority of the world's carbon is in the

non-radioactive form, C12. However, a small percentage of carbon is in the radioactive form, C14. It's this small percentage of C14 which is useful for radioactive dating.

Radioactive decay is measured in terms of half-life. The half-life of C14 is 5,730 years, which means it takes 5,730 years for half of its atoms to decay. For example, an object that contains twenty grams of C14 atoms will only contain ten grams of C14 atoms in 5,730 years' time. Similarly, in another 5,730 years the object will only contain five grams of atoms. This halving process continues for ever.

Although C14 is continually disappearing due to radioactive decay, it is also continually being created. In fact the net result is that in the atmosphere, the ratio of C12 to C14 atoms remains roughly constant. Since the life on our planet breathes this air, this ratio also remains roughly constant in living animals and plants. However, once these creatures die and stop breathing, new supplies of C14 are no longer obtained. From this point on the ratio of C14 to C12 begins to drop. The amount this ratio has dropped below the atmospheric value then determines the age of the object. For example, if the C14 to C12 ratio in an ancient wooden tool is half of that in living trees, the tool must be 5,730 years old.

In the past radio dating fossilised bones and teeth has served anthropologists well and will, no doubt, continue to do so. However, molecular biology is providing a brand-new angle of attack in the search for our origins — an attack that is proving to be very fruitful.

# Canaries can change the weather

*P*erhaps the reason the weather bureau is having so much trouble with their forecasts is because they're unaware that I let my pet canary out to stretch its wings on Tuesdays! Of course I'm only joking, but, as strange as it sounds, it is just possible that the flapping of a canary's wings could have an appreciable effect on future weather. The reason goes back to a remarkable discovery made in the early 1960s. However, to understand this, a small digression is necessary.

The timeless occupation of predicting the future is usually associated with gypsies and crystal balls. However, forecasting future events is very much what science is all about. In fact one of the best tests of a scientific theory is its ability to predict some future event. For example, the ability to predict total eclipses of the sun is powerful evidence that Newton's theory of planetary motion is correct.

However, the idea that the future can be predicted by science runs much deeper than this. In fact a famous eighteenth-century scientist, Pierre Laplace, once remarked that if he was given enough information about the present he could predict the future in every detail and as far into time as he desired. Although this seems like a very bold statement, Laplace had two good reasons for making it. The first is the fact that Nature is governed by a series of laws and these laws are never broken. For example, each time a stone is thrown or an apple falls, the motion is governed by the laws of gravity. In fact the whole of science is based on the premise that there are unbreakable laws in Nature.

The second reason was that, in the eighteenth century, Nature also appeared to be completely predictable. A very simple example is the firing of a cannon at a distant target. If there is no wind blowing, once the angle of the cannon and the speed of the ball are set, simple equations predict exactly where the ball will land. No matter how many times the cannon is fired, the ball will hit in precisely the same place, provided the angle and speed are the same. Nothing unpredictable can happen. From the motions of the planets to the daily weather, in the eighteenth century it seemed that everything was predictable in just this way. In essence, the future is already contained in the present.

Laplace argued that the reason we cannot predict the future with complete accuracy is because we don't know what is happening in the present in enough detail. Once again firing a cannon provides a good example. The speed and angle of firing can only be set so accurately. They will be slightly different each time the cannon is fired, and therefore the cannon ball will land in a slightly different place each time. The accuracy to which the landing place can be predicted is determined by the accuracy to which the speed and angle are known.

Laplace argued that if we could measure the present with enough accuracy and detail we could predict the future with accuracy. In other words, it is not that the future is unpredictable, it is just that we don't have enough information to predict it. Of course, the amount of information required about the present would be horrendous. No less than the position and speed of every particle in the entire universe would be required!

These ideas about predictability have now been around for a long time and in fact underlie much of scientific thinking. However, in recent years, they have been challenged. New developments indicate that the world may not be as predictable as scientists had thought. Weather prediction is a good example. Since the weather is governed by the laws of physics, it has always been thought that the weather should be, in principle, completely predictable. The only reason weather forecasts are unreliable is that not enough information is known about today's weather. It had always been assumed that these predictions would improve as more input information became available. Eventually the weather forecasters would get it right! However, those hopes were dealt a severe blow in 1963 by a meteorologist named Edward Lorenz.

Lorenz was working with some of the simple equations used in weather prediction. However, although the equations were simple, his discovery was momentous. He discovered that some solutions to these equations could never be predicted, no matter how accurately the input information was known.

The problem is that these equations for the weather are very sensitive to the input information. It's a bit like the classic science fiction problem of going back in time. The minutest of changes can make an enormous difference to future events. For example, if the train on which your parents met was a few minutes early your father might have missed it. The consequences, especially for you, would have been enormous!

The same goes for the weather. Even if all rele-
vant measurements of today's weather were known to
an incredibly high degree of accuracy, there would
still be some very small error. And this small error
would serve as a minute change that could have a
major effect on future weather. Even a disturbance as
small as the flap of a canary's wings could determine
future weather patterns. In fact it turns out that no
matter how accurately today's weather is known, the
minutest of measurement errors will grow so rapidly
that future weather could never be predicted.

Equations that have solutions that are very sensitive to input values are termed chaotic. Fortunately
not all equations have chaotic solutions. The
equations that govern the solar system or a cannon
ball, for example, do not. This is why useful predictions can be made with them. In fact the equations for
the solar system can be used to predict eclipses for
hundreds of years in advance. However, scientists are
discovering that more and more of the world is
chaotic. It seems that much of the future may really be
inherently unpredictable.

# Radiation — should we worry?

*E*very minute of our lives our bodies are bombarded by radiation. Radiation from space, from the bricks in houses, from the air, from dentists' X-rays, from nuclear tests, from television sets, from power lines, the list goes on. Even food and drink emit potentially deadly radiation. Some, like dentists' X-rays, we expose ourselves to willingly. Others, such as nuclear fall-out and cosmic rays, we have little choice in. Some protest when high-voltage power lines are put up near their houses and still others are frightened by the microwave oven. The big question is: which ones should we worry about?

Our dentists assure us that X-rays are quite safe, but they insist on leaving the room while they are taken. Scientists assured people that radiation levels in the UK from the Chernobyl accident were quite safe. Yet in the same breath they said there would be British deaths as a result. It's not surprising that confusion and paranoia abound when it comes to discussing radiation.

One of the biggest misconceptions is that radiation is a relatively recent, man-made phenomenon. However, by far our greatest exposure to dangerous radiation comes from natural sources, such as the air and rocks. Mankind has been exposed to these for as long as we have been around. And although they are natural, they are just as potentially harmful as the radiation from a nuclear explosion.

In order to understand how dangerous radiation is, it is important to understand the differences between the various forms. Most radiation is a form of light. Like ripples on a pond or sound through the air, light travels in waves. The distance between the crests and troughs of the waves determines the colour of the beam. For example, imagine it was possible to take a light beam and 'stretch' it. As the beam stretches, the distance between crests and troughs (the wavelength) increases, causing it to change colour. It assumes each colour of the rainbow in turn, from green to yellow to orange and eventually red. As the beam is stretched further it eventually becomes invisible as its colour becomes infrared. If it is stretched further still, the beam changes into microwaves and eventually radio waves.

If this light beam was now compressed it would retrace its steps as the distance between crests and troughs became smaller. Eventually the beam would return to blue. If the compression was continued the colour would change to indigo, then violet and eventually it would become invisible as it became ultraviolet. Further compression would change the beam into X-rays and then gamma rays. In other words, all these different forms of radiation are just different colours of light. It is just that the human eye has only been tuned to the part of the spectrum we call visible light.

It is the shorter wavelength radiation that is known to be dangerous. The reason is that radiation with a short wavelength has high energy and the greater the energy, the more damage the radiation can do to the delicate cells in the human body. It is this damage to the cell structure that can ultimately lead to cancer.

The radiation from house bricks, food, X-rays, nuclear reactors and bombs, for example, are all forms of high-energy radiation. For this reason it is well known that they can cause cancer. However, the chance of contracting cancer from small doses is very remote. And our typical lifetime exposure to natural and man-made radiation is very low. However, if you have enough X-rays, for example, the odds can eventually catch up with you. In the same way, tossing a coin and getting ten heads in a row is quite unlikely. But if you toss a coin long enough you are eventually bound to get a run of ten heads. Dentists leave the room during X-rays to avoid the odds catching up with them.

Because there is an extremely remote chance that even the lowest levels of radiation can lead to cancer, deciding a safe level of radiation is really an arbitrary choice. It's a bit like trying to decide how many times you can safely cross the road. The more times you cross, the greater the chance that one day you will be hit. But it is impossible to decide just what is a safe number of crossings. The same applies to radiation. No radiation level is safe. Natural radiation kills people and any increase kills more. But the chances of contracting cancer from low levels of radiation is still small.

Microwaves and radiation from power lines, however, are quite different. They do not have enough energy to damage cells directly in this way. In fact it is not known whether small doses of this low-energy radiation are even dangerous. However, there have been some studies linking power-line radiation and microwaves with cancer. It may be that these forms of radiation can affect the body in some other way.

One study conducted in Poland found that soldiers who were regularly exposed to heavy doses of microwaves contracted cancer at three times the rate of other soldiers. However, the results are far from conclusive. The cancer may have been caused by other common factors or the results themselves may have just been a coincidence. Perhaps the same soldiers were also exposed to some other cancer-causing substance. Or it could be that the soldiers were accidentally exposed to a single enormous dose of microwaves at some stage in their lives and this caused all the damage.

Research in this area is only at a preliminary stage. More studies need to be done before an answer can be given one way or the other. Of course, there is no danger using a microwave oven. Microwave ovens are sealed so the microwaves do not escape.

# Catch an elevator into space

*I*magine taking an elevator into space. Stepping in at ground zero and getting out 36,000 kilometres above the Earth's surface! Or, if that doesn't take your fancy, what about racing through space aboard a cosmic railway moving on tracks suspended tens of thousands of kilometres above the Earth? Although at this stage these ideas are only dreams, they should be taken seriously, for they are the dreams of space scientists.

Space exploration is one of mankind's greatest technological achievements. With the exploration of the outer solar system well under way and queues starting to form for joy-rides on the space shuttle, Neil Armstrong's first steps on the Moon seem a long time ago. However, there is little doubt that the most exciting era of the space age is still to come. And there isn't a better way of getting a glimpse of what might be in store than through the dreams of space scientists. In fact already somewhat less grand versions of the space

69

elevator and the cosmic train are receiving serious attention in space research centres around the world.

Paul Birch is a space scientist with big dreams. He has come up with the concept of an enormous ring in space that would circle the globe. His idea involves putting a series of hollow magnetic coils into orbit around the Earth. Once in place, these coils could be formed into a massive hollow tube circling the entire Earth. A continuous loop of conducting wire could then be placed in the tube and, through a combination of electromagnetic and mechanical forces, the ring would become a rigid structure.

A rigid ring in space would have many uses. For example, it could serve as a dock for spacecraft or as a halfway point for launching rockets into space. However, one of the more exciting possibilities is that it could serve as a cosmic railway line, ferrying passengers from one space station to another.

Other, less grandiose forms of a rigid ring in space are probably more feasible at this time. Such a ring does not necessarily have to encircle the entire Earth. An arc a few hundred kilometres long could be suspended in space.

Another even more ambitious dream is the colonisation of Mars. Although an old one, this dream has received serious consideration in recent years. One of the biggest obstacles preventing a human colony living on Mars is that the Martian atmosphere is unbreathable. Rather than being rich in life-giving oxygen, the atmosphere is largely composed of carbon dioxide. However, scientists are now thinking seriously of ways to change this. One promising idea is to dump tonnes of blue-green algae on to the planet. Over thousands of years this algae would slowly convert the carbon dioxide in the atmosphere into oxygen. After all the Earth itself originally had an atmosphere rich in carbon dioxide. Blue-green algae were responsible for creating the oxygen on Earth

today. Robert Bender from the University of Michigan believes this project is not only feasible but also that it could be done in our lifetime.

The dream of a space elevator is also a dream shared by many. It was partly developed by Jerome Pearson from the US Air Force Flight Dynamics Laboratory. He says that, despite their awesomeness, space elevators are actually based on very simple physical principles. No new scientific developments are required to make them possible. It is purely an exercise in engineering.

One of the biggest engineering problems is that a structure reaching from Earth to space would suffer tremendous stresses. The weight alone would be enormous. Consequently the structure would have to be built from a material far stronger than anything known today. However, space scientists have been devising ways around this problem. Rather than building an elevator supported on Earth, one solution is to design an elevator that is suspended from space. One of Birch's rigid space rings would be ideal for this purpose. It could easily support a space elevator without the use of super-strong materials.

There would, of course, be many benefits of an Earth-to-space elevator. Dangerous and expensive rocket trips into orbit could be replaced by a simple elevator ride. Elevators would provide a convenient means of transport between a space station and Earth. Space elevators could even be used as a very economical means of launching satellites into orbit. In fact, they are ideal for launching geostationary satellites. These are satellites that orbit the Earth at the same rate as the Earth spins. Consequently they always remain above the same place on Earth. Such satellites act as stationary transmitters, relaying telephone and other data signals.

Geostationary satellites must be placed in special orbits. Their forward speed must be sufficient to stop them falling back to Earth under the influence of gravity, while it must also be such that the satellite completes an orbit every twenty-four hours. It turns out that the height for a geostationary orbit is 36,000 kilometres.

A simple way to launch a geostationary satellite would be to carry it to 36,000 kilometres in the lift and then nudge it out the door. Since the Earth spins once every twenty-four hours, a space lift attached to Earth would also complete an orbit every twenty-four hours. And if the tower is orbiting once every twenty-four hours, the satellite will do the same, just as a stone dropped from a moving car automatically has the speed of the car.

However, space elevators are not only useful for geostationary orbits. They also provide an economical way of launching any spacecraft by simply exploiting the physics of gravity and rotation. While at this stage a full ground-to-space elevator remains little more than a dream, smaller elevators suspended in space are being considered.

# Can we contact our cosmic cousins?

*Y*ou might think it's a curious fact that most astronomers actually do believe that intelligent life exists elsewhere in the universe. Curious, because it seems that astronomers are the first to denounce the many UFO sightings reported each year. Even more surprising, perhaps, is that astronomers have also tried to make contact with extraterrestrial life.

Of course it's not that scientists don't believe that UFO sightings are made, they simply don't believe that such sightings provide evidence of visitors from space. Instead, they say that all UFO sightings have perfectly normal explanations. Weather balloons or aeroplanes seen under unusual conditions, for example, have all been reported as unidentified flying objects. One of the big problems is that most people are not used to looking at the sky and consequently mistakes are easily made.

UFO enthusiasts, of course, don't agree. They say that there are many reported sightings that simply defy explanation. And that scientists will do anything to explain away sightings, no matter how unlikely the explanation. The debate between the two camps has been long and bitter and often reduced to little more than a trade of insults. In fact it seems that the only two points they agree on is the existence of extra-terrestrial life and the desire to make contact with it.

It was November 1974 that saw United States scientists make an attempt to contact our cosmic cousins. They broadcast a radio message into space. A large radio telescope was directed towards a cluster of stars about 200 thousand trillion kilometres away, in the constellation of Hercules. The hope was that some alien being somewhere in that direction might tune in. The message, which contained information about us, our planet and our solar system, is still winging its way through space. Perhaps it will never be received or, even worse, received but not understood. But at least an attempt has been made. However, even if a reply is sent it will be a long time before we receive it. Our message will take about 24,000 years to reach its destination and a reply will take just as long to return. It may take 50,000 years for a two-way conversation to be set up!

The reason many scientists are confident that intelligent life exists elsewhere comes from observing the universe. Looking at other stars shows that there is nothing very special about our star, the Sun. There are many other suns just like ours throughout the universe. Perhaps some of these also have planets and if so, there may be many just like the Earth. Since intelligent life evolved here, why not on other Earths?

Of course, one of the keys in this argument is the question of just how common stars with planets are. So far we only know of one for sure, the Sun. However, evidence for other planetary systems is mounting. Using telescopes to 'see' planets orbiting distant stars poses problems. The reason is that stars are extremely bright while planets are dull. Trying to see planets surrounding stars is like trying to see a lighthouse keeper standing next to a searchlight.

However, although planets can't be seen, their effects can be measured. A star surrounded by orbiting planets wobbles slightly due to the gravitational pull of each of the planets. And this wobble can be measured. In 1988 a team of Canadian astronomers observed a number of stars, looking for signs of orbiting planets. They found that over half the stars they observed showed some signs of a wobble. Planets may actually be very common in the universe.

Further evidence of other planetary systems comes from a telescope that was launched into space in 1983. This telescope can 'see' infrared radiation. To the surprise of many, it found infrared radiation to be coming from several nearby stars. The radiation

seemed to be due to a disc of relatively cool matter surrounding the star. This disc, probably composed of small solid particles, is likely to be a planetary system in formation.

In addition to radio broadcasts, other attempts at extraterrestrial communication have been made. Long-playing records carrying a recorded message were placed inside the two Voyager spacecraft that left Earth in 1977. The Voyager project has been extremely successful, the two rockets having explored many of the planets in our solar system. However, once that job is finished the two explorers won't be returning home. They will continue through un-charted space indefinitely. Hopefully, at some time during their travels they will be discovered by another civilisation who will read the messages contained within them.

As well as sending messages, astronomers have also tried listening for them. If intelligent life does exist somewhere out there, perhaps they are trying to do the same thing as us. Such a civilisation may even be far more advanced than us and have an elaborate communication system in place in space. If so we may be able to do some interstellar eavesdropping. Unfortunately, no alien broadcasts have been found as yet.

Although communication attempts have only started relatively recently, we have actually been send-ing messages for the last sixty years. Inadvertently we have been broadcasting our existence into space since the beginnings of radio. Because radio waves travel at the speed of light and light travels about 300 thousand kilometres every second, our earliest broadcasts would be now just reaching places 600 trillion kilo-metres from Earth. As each second passes these pion-eering broadcasts penetrate a further 300 thousand kilometres, advertising our existence to all and sundry.

However, some see advertising our existence as a mistake. Just as Aboriginal life and culture were overwhelmed after discovery by Captain Cook, Earth life and culture may suffer a similar fate, once discovered. Perhaps the Earth should remain nature's best kept secret.

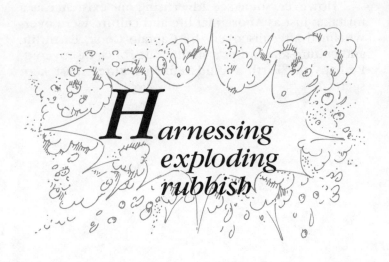

# Harnessing exploding rubbish

Spontaneous combustion is a phenomenon whereby objects catch fire for no immediately obvious reason. Haystacks are a good example. A bale of hay, away from any flames or external sources of heat, can suddenly burst into flames. The cause is bacteria. If the centre of the bale is moist, bacteria can grow happily and heat is produced in their fermentation process. If this heat cannot escape, the temperature rapidly rises and the bale eventually spontaneously bursts into flame.

Another side-effect of bacterial decay is the production of gas. One of the dangers at a rubbish dump, for example, is methane gas. Methane is produced as the rubbish decomposes in the absence of oxygen. If the tip is not properly planned, leaking methane gas can gather and the consequences can be dire. For example, people have died of asphyxiation after becoming trapped in a confined space where the methane has gathered. However, a greater danger is the explosive mixture formed when methane mixes with air. During 1986 in England, a methane explosion at a rubbish tip injured a family of three. The methane escaped from the tip and leaked into a nearby bungalow where it was ignited by the central heating boiler's pilot light. The bungalow was demolished in the explosion.

The methane gas produced at rubbish tips, however, can be put to a useful purpose. It can be harnessed and used as an alternative source of energy. Research into alternative power first became popular during the oil crisis in the 1970s. Then the concern was that the world's reserves of crude oil, our most cherished fossil fuel, looked like they might be running dry. In hindsight, however, the panic turned out to be unnecessary. Today research into alternative energy is again becoming important, but this time it's for a different reason. Environmental pollution is the issue of the late 1980s and will no doubt continue to

be so into the 1990s. One of the big contributors to the greenhouse effect is the carbon dioxide produced by burning fossil fuels. An alternative to oil, gas and coal must be found if the greenhouse effect is to be avoided.

Landfill gas, as the mixture of methane and carbon dioxide produced in rubbish tips is known, is extracted using wells driven into the body of the refuse. Pumps suck the gas to the surface where it is filtered and some of the water removed. It is then ready to serve as an energy source. Its most common use is to generate electricity, although it can be used to fire furnaces or even serve as a fuel for internal combustion engines.

Plants exploiting landfill gas are being set up in many countries. The United States, for example, has more than seventy sites and leads the world in the commercial exploitation of this alternative energy. At one site, in Puente Hills, California, forty-six megawatts are generated and exported to the local electricity producer.

Unfortunately, like fossil fuels, the exploitation of landfill gas also produces carbon dioxide and so does contribute to the greenhouse effect. However, there are other alternative sources of energy that produce no gaseous emissions. Wind power, for example.

The United States, Denmark and Holland lead the world in the exploitation of wind energy. Giant wind farms consisting of many windmills, or more correctly wind turbines, have been built and useful energy has been extracted. However, while power from the wind doesn't produce damaging chemical emissions, there is the potential for other environmental problems. Noise pollution, for example, is a possible concern. However, if wind parks are built far enough from inhabited areas, noise from the spinning turbines is usually not a problem.

Another problem is the sheer size of these modern windmills, which can be 100 metres high. Large wind farms with many turbines could look very ugly. Both the wind turbines and the design of the parks themselves must be carefully planned to minimise their unsightliness. The use of slender blades for the turbines, for example, helps make them more difficult to see from a distance.

Another problem with wind turbines is that they can interfere with radio transmissions. One way of reducing this problem is to make the blades from non-metallic substances. However, this does push up the cost of the machine. Another possibility is to install a booster relay transmitter for the radio signals.

As we move into the next century, wind power will be one of the most important forms of alternative energy. Of course the exploitation of wind energy is best suited to very windy places, although once it is generated, the energy can be transported to other parts. In fact one of the dilemmas facing the wind energy planner is just where to place the wind farms. On one hand, it is best to spread the farms all over the country so that the wind is bound to be blowing past some of them at any time. But on the other hand, the higher the wind speed, the more energy generated. In fact, for a doubling of the wind speed, the amount of energy generated goes up eight times. This is a strong argument for placing all the wind farms in the windiest areas. Fortunately, one of the advantages of wind energy is that if the turbines are put in the wrong place, they can be moved to another location with relative ease.

There are many other possibilities for alternative energy. Energy from the tides, geothermal energy, wave energy, solar energy and, of course, nuclear energy are some of them. They each have their advantages and problems. They are examined next, in *Shooting the tidal rapids*.

# *S*hooting the tidal rapids

*I*n the north-west of Australia, the tide doesn't come in, it pours in. It rushes through the gorges and inlets like a raging river, complete with rapids and whirlpools. The cause is the enormous tidal range in these parts. In just six hours the tide can rise twelve metres, a fearsome statistic considering that more typical tidal ranges are less than two metres.

Watching a rubber dinghy shooting the rapids gives a vivid impression of the energy trapped in these tidal movements that, if harnessed properly, could provide a sensible alternative to fossil fuels for at least some power requirements.

The largest tidal power plant in the world is at the Rance estuary in France. It's capable of generating 240 megawatts of power, roughly enough energy for a town of 100,000 people. Other countries have also built or plan to build tidal power stations across some of their estuaries. The main requirement, of course, is an estuary with a large tidal fluctuation.

Tidal fluctuations can vary markedly from country to country and even within a country. The largest in the world occurs in the Bay of Fundy in Canada, where the difference between low and high water can be more than twelve metres. On the other hand, in the Mediterranean the tidal range is less than 0.6 metres.

The tidal energy is extracted by building a barrage of turbines across the entrance of the estuary. As the tide runs in and out, the rushing water turns the turbines which in turn generate electricity. The technique is similar to the generation of hydroelectric power, where gushing rivers are diverted through a series of turbines.

On the surface, power generation from the tides seems like an environmentally positive alternative. There are, after all, no harmful chemical emissions. However, tidal power stations can damage the environment in other ways. The problem is that the system of turbines needed to extract the energy alters the flow of water in the estuary. A barrage across the entrance damps the flow of the tide. Low tide is no longer as low and high tide not as high after a tidal power plant is built. This means that some of the mud flats in the estuary dry out while others stay permanently flooded. This can be catastrophic for the wildlife in the area.

Another environmental problem for tidal power is that the barrages can also lead to an accumulation of pollutants in the estuary. A barrage of turbines reduces the flow of water in and out and so the natural flushing effect is reduced.

Another source of energy from the sea is waves. Hawaii, the surfing capital of the world, bears witness to the power locked up in a heavy surf. During a fierce storm, the pounding waves can be big enough to tear away large chunks of coral and dump them on the beach. However, the exploitation of wave energy has, so far, posed some problems. It is still very much in the experimental stage. Engineers have tried a number of different ways of extracting the energy but none has been outstanding. As with tidal energy, the idea is to convert wave energy into electricity.

The more usual way to generate electricity is with hot water and steam. In a conventional power station, coal is burned to heat water to a point where it becomes energetic enough to turn electricity-generating turbines. In a nuclear reactor, nuclear fuel is used to do the same thing. In principle, any source of heat strong enough to heat water to high enough temperatures can be used to generate electricity. Such a heat source can be found in abundance under the ground.

Although on the surface our planet is relatively cool, not far below things start to heat up. In fact on average the temperature increases with depth by thirty degrees Celsius every kilometre. This means that at a depth of only four kilometres, the temperature is already above the boiling point of water. The heat is due mainly to the radioactive decay of uranium and other elements in the rocks below. And the high temperatures have given rise to vast underground reservoirs of super-heated water and steam. Harnessing the energy trapped in these reservoirs is the aim of a geothermal power station.

The first modern geothermal power station was built on the Kamchatka peninsula in the far east of the Soviet Union in 1967. It is capable of generating eleven megawatts of power, the energy requirements of approximately a thousand homes. The energy is extracted from a large underground reservoir of steam, which the Soviets believe is only a tiny fraction of their country's total underground steam reserves. They estimate their total geothermal potential to be more than 150,000 megawatts, with the figure likely to be revised upwards.

However, for countries without huge subterranean steam reserves, geothermal energy may still be a possibility. Experimenters around the world are now working on a new way to mine the Earth's heat. Rather than tapping into a natural steam reservoir, they are creating their own steam in hot dry rocks.

The idea is to drill a well shaped a little like a U. The well forms a loop going several kilometres down into the rock and then back out to the surface. Cold water pumped down one side of the well is heated on its way through the hot rock and extracted as steam at the other end. While the drilling is difficult and expensive, engineers involved in this dry rock geothermal energy believe it has a bright future.

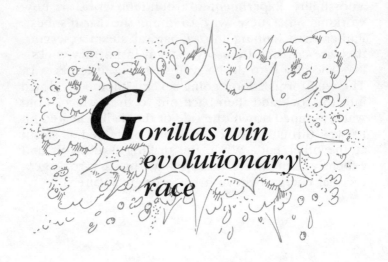

# Gorillas win evolutionary race

*W*hat's the difference between a human being and a caged gorilla? Some would say the cage. However, Roscoe Stanyon at the University of Pisa in Italy would say that the difference lies in evolution. His experiments have found that gorillas have actually evolved further than man!

Mankind's ego suffered a devastating blow with the discovery of evolution. After all, the household dog rose from best friend status to something more like a distant cousin. Even the humble single-celled amoeba could claim a place in our family tree. One consoling point, of course, is that man is truly king of the jungle. We do seem to be on the highest branch of the evolutionary tree. However, the question is: how do we measure evolution? Stanyon's team of researchers measure it by comparing genes. The current thought is that man and ape split away from a common ancestor between five and ten million years ago. Therefore the amount of change in our genes over the last five to ten million years is a measure of how far we have evolved since this time. Stanyon's team found that chimpanzees' and gorillas' genes have actually changed more than man's. In other words, genetically we are closer to our common monkey relatives than either the chimpanzee or the gorilla. Of course, this method takes no account of which genes have changed. And this is an important point. Obviously our few changes have been more beneficial than the greater changes for the apes.

But although these changes are important for us, they must pale into insignificance for Mother Nature. After all, in the beginning, the Earth's only inhabitants were simple creatures. And to call them creatures is being generous. This early life was nothing more than algae. However, in just four billion years Nature has evolved this primitive seaweed into every form of life we see on the planet today.

The process of evolution is truly remarkable, because it is so powerful and yet so simple. One of the

keys to evolution is variety, the fact that no two creatures are exactly alike. Just as every human is unique
so is every animal, bird, fish and even vegetable. From
this large variety of creatures Nature selects only
the fittest to survive. The process is called natural
selection.

The world of business and free enterprise is a
good example of natural selection in action. Every
year many new businesses are born. From chair
makers to chicken sexers, the variety is tremendous.
However, most of these new businesses fail. They fail
for many different reasons, but the important point is
that the ones that survive are the fittest, the ones best
adapted to the world of business. In the same way
most species on Earth produce a lot more offspring
than can possibly survive, and only those best suited
to life on our planet live.

The giraffe's neck is a good example of evolution.
Over the years giraffes' necks have definitely become
longer. However, this isn't the direct result of stretching for leaves on the high branches. Rather, by some
fluke, a giraffe was born with genes that caused it to
have a longer neck. Now for most of the time an
unusually long neck would be no real advantage. However, in times when food is scarce an extended neck
would have a very definite advantage. A taller giraffe
could obtain food by reaching the higher branches.
During these times many of their shorter-necked com-
panions would perish from starvation. The tall giraffe
passes on its genes to its giraffe children and some of
these children will also have long necks. The children
eventually breed, continuing the process. And in several generations there are large numbers of long-
necked giraffes. In fact the long necks will ultimately
take over. This is because every time there is a food
shortage many of the giraffes with short necks die. The
tall giraffes, however, are largely unaffected. As a
result, today we see giraffes with longer necks.

The important point is that this evolution has taken place because of a random mutation or, in other words, an error in the giraffe's genes — in the genes that caused an exceptionally long neck. Although continual stretching of your neck will lengthen it, this longer neck cannot be passed on to your children. Only genes can be passed on and no amount of stretching can change genes.

It's a humbling thought that all life on Earth has evolved because of random errors in our genes. We are merely the product of a long series of mistakes! But that's the way it is. And, after all, we humans make things in a similar way. We call it trial and error. The key point is that by making genetic mistakes Nature can try something new. Even though the vast majority of mistakes are useless and probably result in the death of the animal, there are a few successes. And these are the ones that count.

Changes or mutation of genes can occur for a number of reasons. Sometimes our bodies make a mistake and a gene is accidentally changed. Some chemicals can cause changes in genes. However, one cause of mutation has gained fame in recent years: radiation. Dangerous radiation occurs naturally on the Earth. The Sun, the ground we walk on and even the bricks in our houses are just a few sources of radiation. Fortunately, these natural radiation levels are low. But over the whole population and over a very long time, natural radiation causes mutations. It allows Nature to try something new. However, fall-out from nuclear bombs and reactor accidents raise the level of radiation. The result is that the number of mutations rise. And if the mutation rate is too high life may become extinct before a useful mutation is found.

# *M*ysterious explosion in Siberia

*I*n 1908, in central Siberia, an enormous explosion shook the forest. Two thousand square kilometres of flattened trees and a rising mushroom cloud seemed to point to a nuclear explosion. The problem was that the first atomic bomb wasn't to be tested for another thirty-seven years!

The explosion was devastating. It started a flash fire that burnt thousands of trees and sent a tremendous shock wave through the Earth's atmosphere. A shock wave that actually travelled twice around the globe. In fact the Siberian explosion was so unusual that it is still featured in books and television programs as one of the world's great mysteries. Many explanations have been proposed.

The idea that the explosion was caused by an atomic bomb has many problems, not the least of which is the absence of nuclear fall-out. However, some of the other proposed explanations seem equally bizarre. For instance, some scientists have suggested that the event could be explained by a stray chunk of antimatter crashing into the Earth.

Antimatter, as the name suggests, is the opposite of matter. Everything on Earth is made from ordinary matter, that is, matter composed of ordinary atoms, having a positive nucleus surrounded by negative electrons. However, it is possible to create antiparticles — particles such as positive electrons and negative nuclei. Antimatter is matter built from these antiparticles.

Provided matter and antimatter remain separated, the two can exist quite happily. However, when matter and antimatter come into contact, each completely annihilates the other with a tremendous explosion. Both the matter and antimatter are converted into pure energy.

Although the Earth and its neighbourhood are made from ordinary matter, it has been speculated that there may be regions of the universe that are predominantly made from antimatter. And if antimatter does exist elsewhere, perhaps a chunk of it crashed into central Siberia in 1908. Could this be the explanation for the mysterious explosion?

Although it may seem tempting, there are problems with this theory as well. Although antimatter has been created and observed in the laboratory, the idea that large chunks of it exist elsewhere in the universe is purely speculative. None has been observed. However, more importantly, when matter and antimatter annihilate, there is a strong burst of gamma radiation. No radioactivity has been measured at the impact site.

Another explanation is that the explosion was caused by a mini black hole passing straight through the Earth. This black hole supposedly plunged into the ground in Siberia and exploded out through the North Atlantic later in the day. Once again, the concept of a mini black hole is speculative, but, if such an object did exist, it would wreak havoc if it collided with Earth. Although mini black holes would be very small, their tremendous density would make them lethal. A mini black hole could punch a hole right through the Earth with incredible ease.

However, there are problems with this explanation as well. The records show no signs of an object bursting out through the North Atlantic and, amazingly, there was no impact crater at the site of the explosion. Considering all of the facts, there seems to be only one reasonable explanation for the explosion: on the morning of 30 June 1908, central Siberia was hit by a piece of comet.

Comets are made mostly of ice. If a large piece of comet entered the Earth's atmosphere, a glowing fireball would be seen as it rapidly melted. On impact there would be a massive explosion which could flatten the forest and even cause a rising mushroom cloud. After the explosion there would be no radiation fall-out and there would be no impact crater, as the comet would melt when passing through the Earth's atmosphere. A perfect explanation for the Siberian explosion.

Like the Earth, comets also belong to the Sun. They orbit it in enormous elliptical paths. But unlike the Earth, which takes a year to complete its orbit, comets usually take much longer. Halley's comet, for instance, completes an orbit every seventy-six years. At one end it is out near Pluto, the outer-most planet, while the other end of its orbit takes it very close to the Sun.

The origin of comets is still unknown. Astronomers believe that there is a huge reservoir of them surrounding the Sun with orbits way out beyond the planets. This reservoir, known as the Oort Cloud, contains billions of comets. The theory is that, every so often, the Oort Cloud is shaken up, causing some of these comets to fall out of the cloud and into the inner solar system. It's only after they have entered the inner solar system that we see them as comets. Such a shake-up could be caused by the passing of a nearby star or even by the gravitational pull of the galaxy itself.

For most of its life a comet looks very different from the way we see it. Its characteristically long tail only develops as it passes close to the Sun, and for most of its passage through space it has no tail. The tail consists of gases that have evaporated from the comet as it passes near the Sun and can stretch out for millions of kilometres. Halley's comet's tail, for example, can be thirty million kilometres long.

A comet's tail also disobeys intuition. Looking at a comet as it rushes through space, it's natural to assume that its tail streams out behind it. However, for much of its journey this isn't true. In fact at times a comet travels through space tail first, the reason being that the tail always points away from the Sun, the source of heat. Living on Earth we have become brainwashed into thinking that an object's tail is always behind. This is because everything on Earth is affected by the force of air resistance, which forces the tail to stay behind. However, in space there is no air resistance and therefore no force to push the comet's tail back into line.

# Safe radiation can kill

*I* still have vivid memories of the day I stepped out of the front door into a cloud of nuclear radiation. It's a weird feeling knowing that the air you breathe and the ground you walk over are radioactive. Weird, because everything seems perfectly normal!

I was in England when the radioactive cloud from the ill-fated Chernobyl reactor passed over. Within a few days the media carried warnings not to drink the rainwater. Within a week higher than normal radiation levels were found in cow's milk. And within a month radioactivity had even made its way into the Sunday roast lamb.

The official word was that there was no need for panic. In England the radiation levels were safe. Yet at the same time scientists were predicting English deaths as a result of the radioactive cloud! You can imagine the confusion this created. How can safe levels of radiation kill people? Although it seems hard to believe, radiation levels can be safe even though some people will die. To understand why you must first understand radiation.

Dangerous radiation is a natural part of our environment. Since the dawn of time all life on Earth has been exposed to it — radiation from outer space, radiation from the ground we walk on, even radiation from the bricks in our houses. In fact the human body itself is radioactive! Exposure to this natural radiation is unavoidable and, like all dangerous radiation, it can cause cancer and mutations. People die as a result of Mother Nature's radioactivity.

However, mankind's efforts can drastically increase the level of radiation we are exposed to. The level of exposure was increased for all those who lay in the path of the Chernobyl cloud.

Trying to decide a safe level of radiation is a bit like trying to decide whether long weekends are safe. Every weekend people are killed on the roads. Every long weekend more people are killed. Although the chances of being killed are greater on a long weekend, they are still small. After all, if they weren't none of us would bother driving.

The same applies for radiation. No radiation level is safe. Natural radiation kills people and any increase kills more. But the chances of getting cancer from low levels of radiation are small. Just as the road toll is higher on long weekends, the cancer rate will be higher after the Chernobyl accident. But my chances of contracting cancer are still remote.

However, England is a long way from Chernobyl. Closer to the accident a lot of people were exposed to high levels of radiation, and the effects of this will not be known for many years.

Nuclear technology was dealt a severe blow by the Chernobyl accident. Many nuclear supporters must have had to rethink their stance. However, the idea of a nuclear future still has a strong appeal. In fact scientists envisage nuclear power of the future to be very different from that of the present. They are trying to develop reactors that produce no radioactive waste and use water rather than uranium as nuclear fuel.

Nuclear energy is the energy of matter itself. In a nuclear reaction matter is destroyed and converted into pure energy. However, there are two very different ways of extracting this energy. One method is to fuse a number of atoms together, called nuclear fusion. The other is to split atoms apart, called nuclear fission. Either method produces nuclear energy but nuclear fission, or atom splitting, is a lot easier to achieve. The first atomic bombs and all present-day nuclear reactors derive their tremendous energy from atom splitting.

A major problem with extracting energy from an atom by splitting it, however, is that the leftover pieces are highly radioactive. This highly radioactive nuclear waste must be stored for many years before it has decayed to a sufficiently safe level.

However, for nuclear fusion, the problem of radioactive waste disappears. Hydrogen atoms, for example, can fuse to form helium atoms and produce energy without the problem of radioactive leftovers. While hydrogen gas is relatively rare on earth, hydrogen is extremely abundant in other forms. Sea water, for example, could be used to satisfy mankind's demands for many many years.

While nuclear fusion is an obvious solution to our current energy problems, producing a controlled fusion reaction has so far proved impossible. A fusion reaction is self-sustaining. Once some atoms have fused, the resultant energy release causes other atoms to fuse, starting a chain reaction. The hydrogen bomb derives its energy in this way. However, rather than being controlled, the fusion process in a hydrogen bomb is uncontrolled. In fact the chain reaction is started by using a conventional fission bomb.

The problem with fusion is to start that chain reaction. Obviously a fission bomb would not be an ideal way to start the process in a nuclear reactor! Fusion scientists are working on other possibilities for triggering the chain reaction.

Three conditions must be met in order to get atoms to fuse. The first is that the temperature of the fuel (hydrogen) must be at least 100 million degrees. The second is that the fuel must be compressed so that there are between 100 and 200 billion billion particles per cubic metre. And the third is that these conditions must be held for at least one second.

Achieving these extreme conditions is no easy feat. These temperatures are greater than those at the centre of the Sun! However, in a recent series of experiments conducted by European scientists, each of these three magic numbers was satisfied. This proved that the conditions needed for a fusion reactor were possible. However, as yet all three have not been achieved at the same time. One of the problems is that the techniques used to increase one usually result in the decrease of the other two.

There are many technical problems that have been overcome to get fusion research to where it is today. For example, simply confining such a super-hot fuel is a difficult task. After all, 100 million degree temperatures would instantly disintegrate any known physical substance! Instead of using a physical container, the hot fuel is held in place by a magnetic field.

Although the problems have been great and there is still more work to be done, progress is being made.

# The heat is on . . . or is it?

*T*here is a world not far from Earth where the climate is atrocious. Night-time temperatures plunge to −150°C while those during the day are about 100°C above zero. This world is the Moon. In fact the average temperature on the Moon is around −18°C, compared to a relatively warm +15°C on Earth.

At first this may seem surprising since the Moon and Earth are almost exactly the same distance from the Sun. However, the difference is the Earth's atmosphere. This layer of air surrounding our planet acts like a blanket and moderates the climate, keeping average temperatures around +15°C instead of the −18°C they would be without the atmosphere. The warming due to this atmosphere blanket is called the greenhouse effect.

The greenhouse effect is a completely natural feature of our planet and has been operating for as long as the atmosphere has been around. The problem, of course, is that modern man's activities may be

**98**

affecting the natural balance. Industrial man has been adding many gases to the atmosphere, some of which, such as carbon dioxide, methane and chlorofluoro-carbons, enhance the greenhouse effect. A significant rise in the amount of these gases in the atmosphere will cause a greater greenhouse warming.

The big question at the moment is whether this greater greenhouse warming is already occurring. The warmest year since records began was 1988 and the 1980s are certain to go down as the warmest decade on record. Not surprisingly, this has caused many to believe that the greenhouse warming is on the increase. However, trying to decide whether this is so is very difficult. Our current record-breaking warm weather is certainly not proof.

The problem is that the Earth's climate is extremely complicated, involving the complex inter-action of many factors. The warm weather of the 1980s could be just a normal variation.

One of the biggest question marks for climate modellers is El Niño. El Niño is an event that occurs about every four years in the tropical Pacific Ocean. Wind and ocean currents that normally flow from east to west suddenly reverse for a few months, disrupting birds, fish and even the climate. During an El Niño event the normally arid western coast of South America suffers torrential rain while normally wet parts of Australia can suffer drought. El Niño is also associated with the failure of the Indian monsoon and is linked with droughts in Africa.

El Niño also seems to cause a general warming right around the globe. In fact the three warmest years on record — 1988, 1987 and 1983 — all coincided with El Niño events. It could be that the record high temperatures of the 1980s were primarily due to El Niño rather than to greenhouse warming.

This shows some of the problems involved when making predictions about the greenhouse effect. The question in the greenhouse debate is not whether the atmospheric concentration of greenhouse gases like carbon dioxide is increasing — they certainly are — but, rather, whether this increase is causing a rise in average temperature.

The greenhouse effect has received enormous media attention in recent times. However, scientists have known about it for many years. John Tyndall first wrote about greenhouse warming in 1863. In fact, as early as the 1890s scientists were considering some of the problems that could arise if the amount of carbon dioxide in the atmosphere were to increase due to the combustion of coal. However, it was the 1970s that saw renewed interest. By this time an appreciable amount of carbon dioxide had built up in the atmosphere. Greenhouse warming suddenly became a major topic for research.

Some of the initial predictions were alarming. As the Earth warmed, some scientists were predicting sea-levels around the world to rise by many metres. In fact during warmer times in the past sea-levels were much higher. The reasoning was that as the Earth warmed some of the polar ice would melt, causing sea-levels to rise. However, today climatologists don't believe temperatures will rise by enough to melt that much ice and they say there will only be a slight rise in sea-levels.

However, a small rise in the average temperature around the globe could cause other problems. For instance, on a warmer Earth rainfall patterns will change. Although this would be an improvement for some parts of the world, it would be devastating for others. The wheat-growing regions in the United States are a prime example. They supply a great deal of the world with food. If this part of the globe suffered extended periods of drought the effects could be immense.

The reason greenhouse gases such as carbon dioxide cause warming is because they treat heat and visible light differently. Carbon dioxide is transparent to visible light but not to heat. Most of the Sun's energy comes to Earth in the form of visible light. This means that sunshine can pass through the atmosphere and heat the Earth. However, the heat generated cannot radiate back out again so easily. Instead the atmosphere absorbs some of this heat and re-radiates some of it back to Earth. In this way our planet is kept warmer.

Just how much our burning of fossil fuels and cutting down forests will change the climate in the future is not known with certainty. It is vital that more research be done now to avoid potentially massive problems in the future.

# Madness

*M*adness. It's the thing many people fear most — the possibility that a perfectly normal and happy life can quickly and uncontrollably change into one of confusion and paranoia. Schizophrenia has occurred throughout history, in every society around the world. Overall, it affects about one per cent of the population, although in some parts the percentage is even higher.

In the past, the treatment of those suffering schizophrenia has been terrible. Sufferers were usually thought to be possessed by demons and spirits and consequently were subjected to a variety of 'cures'. Remedies included burning at the stake, being chained in dungeons or drowning. Fortunately today we know a little better and schizophrenia is seen as a disease of the mind, although it's still far from completely curable. Psychiatrists say that among those that succumb to the condition, only about one-third will never fall ill again, another third will suffer intermittently for the rest of their lives, and the final third remain permanently ill.

Anyone who has known someone suffering schizophrenia knows just how bizarre and offputting its effect can be. Jumbled conversation, sentences that don't make sense, voices in the head, inappropriate moods and grandiose delusions are all part of a schizophrenic's profile. Witnessing someone actually crossing from sanity to madness can be soul-destroying. And it's made even worse by the fact that the causes of schizophrenia are still largely unknown. While there is increasing evidence that schizophrenia may be genetically inherited, there is also evidence that the environment and experiences of a schizophrenic also play an important role. Trying to sort out which of these influences is predominant is no easy task.

One of the most distressing theories advocating that environment and experiences are the dominant influence became popular in the 1950s. Theodore Lidz from Yale University advanced the idea that certain types of mothers were responsible for their schizophrenic children. These were supposed to be mothers who continually placed their children in situations of conflict where, no matter what the child did, the result would always be a clash. It goes without saying that this theory caused a great deal of stress for the families involved. Not only had a mother lost her child to the disease, but also she was to blame. Fortunately, evidence was never found to support this theory and it was eventually discarded. In fact these days there is increasing evidence that genes play an important role in schizophrenia.

One of the big problems with trying to sort what is caused by genes from what is caused by environment is that children acquire a lot more than just their parents' genes. Parents also provide a child's learning environment. Children pick up many of their parents' habits, their eating patterns, their morals and social behaviour. Simply studying the families of schizo-phrenics cannot separate these two influences.

Instead, a number of special family situations are studied. For example, one way to separate genetic influences from environmental ones is to study famil-ies in which the children have been adopted. Adopted children, in effect, have two sets of parents. One set gave them their genes and another set shaped their environment and experiences. By comparing both sets of parents to the children, some idea of the relative importance of environment versus genes can be gained.

The ideal subjects for these studies are identical twins that were adopted by separate households at birth. Since these children have 'identical' genes but different environments, they offer a chance to see which is most important in causing schizophrenia. However, although some studies have been done, the number of identical twins suffering schizophrenia as well as being adopted by separate households is very small. There aren't enough for a reliable study. Instead, researchers must make do with studies of adopted children generally.

Another alternative doesn't rely on adoption at all. These studies compare the incidence of schizo-phrenia in identical twins to the incidence of the dis-ease in non-identical twins. Unlike identical twins who have the same genes, non-identical twins only share about half of their genes. Therefore, if a higher incidence of schizophrenia is found among identical twins, it indicates that schizophrenia may have genetic causes.

During the last twenty years or so, there have been many studies implicating genes in schizophrenia. However, these studies have been criticised on many grounds. Unfortunately, they do contain possible sources of error. For a start, even diagnosing the dis-ease is a problem. The question of whether a patient is a schizophrenic does not have a clear-cut answer. Schizophrenics are diagnosed on a wide variety of symptoms many of which appear in some but not in others. The diagnosis must eventually involve a subjective element. In fact one of the questions a patient is asked is whether there is any schizophrenia in the family. If there is, the doctor is more likely to diagnose the patient as schizophrenic. This obviously causes bias towards the importance of genetic factors.

Another problem is that adoptions rarely take place right from birth. The environmental influence during very early life, before the adoption takes place, may be important. Further bias can also occur because adoption often takes place between matched socio-economic groups. This means that the adopted home environment is often similar to the environment of the biological parents.

# $S$uperlight

$I$t is a beam of light with so much power that it
can blast through diamond. Yet so controlled, it can
cut delicate human tissue and even play music. The
laser beam has come a long way from its fictional
beginnings as the 'death ray' of cartoon comic strips.
In fact lasers are infiltrating daily life at a staggering
rate, from factories, where they are used to cut anything from sheet steel to clothing fabric, to compact-
disc players and operating theatres. However, despite
their tremendous power and high-precision applications, a laser beam is simply a beam of light.

Unlike the light beam from a torch, however, a
laser beam has a number of special properties. For a
start, a laser is a highly directional beam of light. A
beam that is one centimetre wide when it leaves the
laser, for example, may only spread to ten centimetres
after travelling for a few kilometres. This ability to
emit light in one direction is unusual. Most light
sources radiate light in all directions. The filament in

a torch globe, for example, emits light in all directions. To form a torch beam, this light must be focused by a curved mirror just behind the globe. But for lasers the light beam itself is only emitted in one direction. The consequences of such a concentrated beam of light are great.

Because a standard 100-watt light globe sends out light in all directions, the power that actually enters the eye when you look at it is very small — less than one-thousandth of a watt if you're about thirty centimetres away. However, if you looked into the beam of a 100-watt laser all one hundred watts would enter your eye. In other words, a 100-watt laser is about 100,000 times more powerful than a 100-watt globe! Of course lasers can be made that deliver much more power than 100 watts, and in fact lasers can be made strong enough to vaporise any known material.

A second unique feature of the laser is that it produces light of a very pure colour. Most light is not at all pure. Sunlight, for instance, is made up of every colour in the rainbow. After all, rainbows form as water droplets split sunlight into its constituent colours. However, even the most pure colours our eyes see are relatively impure. A red car, for instance, is actually many different shades of red. It's just that the difference between the shades is so small the human eye cannot detect it. But in contrast the red light beam from a ruby laser is very pure. It only consists of a single shade of red.

Another property of laser light which makes it different from other light sources is coherence. To understand coherence it is important to know something about light. Light travels in waves, like ripples in a pond or sound through the air. But unlike the crests and troughs of water waves which may be metres apart, the crests and troughs of light waves are extremely close together — less than a thousandth of a millimetre apart. This distance between the crests and troughs, or the wavelength, determines the colour of the light. Red light, for example, has a larger wavelength than blue light.

However, not all light is visible. Ultraviolet radiation, for example, although invisible is just another 'colour' of light with a wavelength shorter than that of visible light. Similarly, radio waves are a 'colour' of light whose wavelength is much greater than that of visible light. It is just that the human eye has only been tuned to see in the range we call visible light.

A laser makes use of fluorescent substances, substances that actually give off their own light. Of course they can't generate their own light without a power source and what they do is absorb light of one colour and re-emit this light as another colour. However, the light they absorb does not need to be visible. In fact some substances absorb ultraviolet light, which is invisible to the human eye, and re-emit it as visible light. These objects appear to glow in the dark.

It is the atoms that make up the object that absorb and re-emit the light. Normally each atom emits light whenever it pleases, with no regard for what the other atoms are doing. Since every object is made up of millions of atoms, the result is that the light given off is a jumbled mess of crests and troughs, just like the ripples that form on a lake when a handful of stones is thrown in. Such a jumbled mess isn't able to exert much force. The situation is a little like a group of soldiers crossing a bridge. Provided the soldiers break step, their marching exerts little force and they can cross the bridge with relative ease. However, if they march across in step, the force of hundreds of boots crashing down together can cause the bridge to collapse.

A laser has the effect of coordinating the atoms so that they emit light at just the right time and the crests and troughs of the light emitted from one atom match up with those from another. The result is that lasers emit very smooth and powerful light waves. Like soldiers marching in step, the crests and troughs from each atom are 'pushing' together.

In a fairly new application, lasers are one of the latest weapons being tried in the fight against cancer. However, rather than being used as a cutting instrument, lasers provide a source of light that is used in combination with chemicals. These chemicals, named porphyrins, have two special properties. The first is that if they are injected into a human or animal body they tend to concentrate in tumours. The second is that when these chemicals are exposed to laser light, a chemical change occurs which causes the cancerous cells in the tumour to die. Even when the tumour is deep within the body laser light can be passed down an optical fibre and directed into the tumour.

While at this stage the technique is still very new, scientists testing it say its future looks promising. They believe that in the not-too-distant future it should be a significant weapon in the fight against cancer.

# Gravity — an illusion

*H*igh above the Himalayas one night, I had the chance to experience life without gravity. The familiar 'Fasten seatbelts' sign flashed on as our jumbo was about to encounter 'some turbulence'. As I buckled up, I asked myself why I bothered. A little turbulence could hardly throw me out of my seat. I was wrong. We hit an air pocket and the plane went into a dive. The recently served orange juice headed for the roof and the plastic cups were not far behind. I remember glancing up and seeing a sea of orange juice and cups floating about the overhead lockers. What seemed like minutes later, but was probably only a few seconds, the plane recovered and gravity was restored to normal. The orange juice and cups came down.

Gravity is essential for life on Earth. Without it, our planet and its contents would float off into space. However, Albert Einstein showed that, in a way, the force of gravity is an illusion. Of course, gravity itself is real enough but Einstein showed that it is not a force as we normally imagine them.

The best way to understand this is through an analogy, or a thought experiment, as Einstein liked to call them. Imagine you took a yacht out into the Pacific Ocean and you decided to sail a triangular course. Say each leg of the course is 500 kilometres. Remembering your school trigonometry, this is an equal-sided triangle, so each of the angles must be sixty degrees. With the weather fine and the winds high you set off on the triangular course. At the end of the first 500-kilometre leg you make a sixty-degree turn and continue with the second leg. Finally, after another sixty-degree turn, you complete the triangle and sail back to the point you started from. At least that's what you expect.

However, if you actually did this trip you would soon discover that something was wrong. For, instead of returning to the point you started from, you would

find that the yacht had drifted to the east. Now your navigation had been perfect. You'd allowed for every breath of wind and even the smallest ocean current. So there's no way you could have sailed off course. What then had caused this inexplicable drift to the east? If you didn't know better, you might think that the answer was a mysterious force. A force pulling to the east. However, there is a much simpler explanation.

The reason is that the Earth's surface is not flat, it's curved like the surface of a beach ball. An equal-sided triangle with all angles equal to sixty degrees is only valid on a flat surface. If you draw one on a beach ball you start to see what's happened. Draw the first leg, make a sixty-degree turn and then draw the second. After another sixty-degree turn, the third leg ends up over-shooting the mark. While an equal-sided triangle on a flat surface has sixty-degree angles, the same triangle on a sphere must have each of the angles greater than sixty degrees. So there is no force pulling to the east. It's the fact that the yacht is moving in straight lines over a curved surface that gives the illusion of a force. Similarly, Einstein said that the force we call gravity is nothing more than the effects of a curvature — a curvature of space and time itself. Rather than the Moon being attracted to the Earth by the force of gravity, it is simply following a straight path in a curved space and time. We think of gravity as a force only because we consider the space we live in is flat. The thing that makes space curve is matter. The Earth itself bends the space around it.

Trying to imagine the space we live in as curved is impossible, so don't try. It's impossible for our minds to conceive. Our minds can only conceive two-dimensional surfaces, like a piece of paper, for example. A piece of paper is two-dimensional because it only has two dimensions, length and breadth. There is no height. However, the real world is three-dimensional. Everything has length, breadth and height. It's impossible to even begin to imagine what a three-dimensional surface might look like. But this is not because they can't exist. It's just that we're three-dimensional creatures.

Since Einstein put forward this theory early this century, the evidence supporting it has continually mounted. One of the more startling confirmations came in 1962. The theory predicts that time appears to travel more slowly in places where gravity is stronger. For example, since gravity is weaker at the top of a mountain, people living there age faster than those living at the bottom. Of course, this time difference is absolutely minuscule and nothing you could ever hope to notice. But this is only because the Earth's gravity is relatively weak. In places where gravity is very strong, near some dead stars for instance, the effects are very noticeable. However, although the effects for the Earth's gravity are small, they can be measured. In 1962 a pair of extremely accurate clocks were placed at the top and bottom of a water tower. Just like a mountain, gravity is slightly weaker at the top of the tower, so time there should pass a little faster. Sure enough, the clocks confirmed the prediction.

You might think that the effects of general relativity on Earth are so small they're of no practical importance. However, this is no longer true. To build very accurate navigation systems based on satellites the effects of general relativity must be taken into account. If they're not, errors of several kilometres can occur. As is so often the case in science, what seems to be purely academic when it is discovered turns out to have a very practical importance later on.

However, when it comes to astronomy, the theory had practical importance immediately. In fact one of the early confirmations of general relativity is that it correctly predicted the unusual orbit of the planet Mercury. And when it comes to some of the weirdest objects in the universe, black holes and other dead stars, general relativity is the only way sense can be made of them.

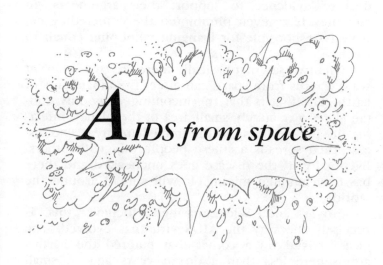

# *A*IDS from space

*T*here are many theories on how, where and when the AIDS virus began. However, one of the most radical would have to be that the virus responsible for AIDS came from space. It made its way to Earth aboard a comet and rained down over the planet as the comet disintegrated in the atmosphere.

The theory is controversial, to say the least, and if it wasn't for the reputations of the scientists involved, it might easily be discarded without further consideration. It is the work of the respected English scientist Sir Fred Hoyle and his colleague Professor Chandra Wickramasinghe. According to their theory AIDS and many other diseases including whooping cough, smallpox, viral meningitis and Legionnaire's disease, to name a few, do not originate on Earth but instead come from space. In fact they claim that even the dreaded yearly bouts of 'flu are attributable to viruses from space.

Although the theory is still in its infancy, Hoyle and Wickramasinghe have already gathered a great deal of evidence to support their arguments. In fact they have even pinpointed the comet they believe is responsible for bringing whooping cough to Earth.

The two say that basically there are two different ways a virus from space can affect the inhabitants of Earth. The first is that the incoming virus only contaminates a relatively small area of the planet, infecting the people living in the immediate area. The disease then spreads as these people mix with others. Eventually, if the disease goes unchecked and there has been enough mixing, it will spread around the world.

Hoyle and Wickramasinghe believe this is precisely the way the AIDS virus has evolved. The virus arrived on a comet that entered the Earth's atmosphere less than a dozen years ago. A small group of people in central Africa were all that were initially infected. However, the disease then spread from person to person through sexual and blood contact.

The other way a virus from space could spread through the human population is much more direct. Rather than only affecting a small area, the comet could spread the virus over large parts of the Earth's surface. Hoyle and Wickramasinghe argue that influenza is a good example. They say that 'flu outbreaks are often reported in cities in different parts of the world simultaneously. The conventional theory for the spread of influenza is that it is transferred by person-to-person contact. However, Hoyle and Wickramasinghe say it is difficult to see how outbreaks that occur simultaneously on different sides of the globe fit in with this theory. Their theory, however, provides a simple explanation. It simply means that the influenza virus rained down over several different cities at the same time. In fact Hoyle and Wickramasinghe report a number of case studies supporting this idea.

One of these is a report written by Dr Louis Weinstein following the most disastrous worldwide 'flu epidemic of recent times. It occurred in 1918–19 and was responsible for about thirty million deaths. After studying all the evidence at hand he concluded that the disease did spread from person to person in local areas, but the disease also occurred in widely separated parts of the world on the same day. 'It was detected in Boston and Bombay on the same day, but took three weeks before it reached New York City, despite the fact that there was considerable travel between the two cities.'

Again similar observations were made after the worldwide epidemic of 1948. A Sardinian doctor, Professor Margrassi, wrote: 'We were able to verify the appearance of influenza in shepherds who were living for a long time alone, in open country, far from any inhabited centre; this occurred almost

contemporaneously with the appearance of influenza in the nearest inhabited centres.'

In their book, *Cosmic Life Force*, Hoyle and Wickramasinghe present further evidence supporting their premise that influenza and other diseases come from space. One of their most fascinating pieces of evidence is provided by outbreaks of whooping cough. It has been known for some time that outbreaks of whooping cough occur approximately every 3½ years. It has always been thought that the reason for this is that most of the susceptible children get the disease during an outbreak and eventually the transmission must stop. It then takes about 3½ years for a new population of susceptible children to develop from birth.

Hoyle and Wickramasinghe claim that the cause of the 3½-year cycle is a comet that visits Earth every 3½ years. During each visit a fresh batch of the whooping cough virus is injected into the atmosphere. The two even go as far as naming the suspect comet. Encke's comet re-enters our atmosphere every 3.3 years and a few simple calculations show that material from Encke's comet would rain down on Earth precisely every 3½ years.

With the idea of putting their theory to the test, Hoyle and Wickramasinghe examined the whooping cough statistics for the period 1960–75. During this time a vaccine for the disease was being administered. For the period that the vaccine was in general use, the conventional wisdom predicted that the number of children infected by whooping cough should decrease and epidemics should become less frequent. In other words, the cycle time should be greater than 3½ years. On the other hand, Hoyle and Wickramasinghe's theory also predicted that the total number of children affected would decrease but that the cycle time would still remain 3½ years. The statistics for 1960–75 certainly show that the total numbers were reduced by the vaccine, but the 3½-year cycle is still clearly present.

Hoyle and Wickramasinghe believe the evidence for diseases from space is enormous. In fact in *Cosmic Life Force* they also argue that life itself did not begin on Earth but came from space. This early space life subsequently evolved into every living creature on the planet today. Perhaps the entire universe is teeming with life!

# *A*n easy bet

*I* am willing to bet a year's wages that you can't tell me the length of Australia's coastline to the nearest thousand kilometres. You might think that this is a pretty safe bet. After all, all you need is a nice big map of Australia and a piece of string. Lay the string carefully around the coastline and then measure the length of the string.

It sounds simple but your answer would be quite wrong. Think about it. What would happen if you took a much larger map and did the same thing? Of course your string would need to be longer but once you took the scale of the map into account you would think that you should end up with the same answer. The problem is that you don't. On the bigger map Autralia's coastline seems to be much longer!

The reason for this is because more detail is shown on a larger map. There are a lot more wiggles and bumps for the string to be laid carefully around. On a smaller map the string just cuts straight across this intricate detail. If an even larger map is used the problem is just repeated. In fact the bigger the map, the longer the coastline!

However, not willing to give up easily, you might naively think there is a way around the problem. Although it would be difficult in practice, in theory you could take an enormous tape measure and actually walk around Australia's coastline taking measurements. You could place your measuring tape around each individual rock and zig-zag of the coastline and come up with what you might think is a very accurate answer. But alas you still haven't succeeded. You can see why by examining the coastline through a magnifying glass. You discover that there is still more intricate detail you have missed. If you repeated the measurements using this magnifying glass you would find the coastline to be even longer still. In fact by simply using stronger and stronger magnifications you could make the coastline as long

as you like! In effect, Australia's coastline is infinitely long.

Now you may think that this is being pedantic. But it certainly isn't. In fact countries often disagree on the length of their common border! It all depends on which scale each country does its measuring.

Obviously this measurement problem causes headaches for geographers. Standard high school geometry just can't deal with shapes that have infinitely long edges. Standard shapes, like squares and triangles, have edges that are easily measured. Their length certainly doesn't depend on the size of the map on which they are drawn. For this reason mathematicians have invented a radically new type of geometrical shape, called the fractal. Fractals are amongst the weirdest shapes imaginable. Just like coastlines, their edges are infinitely long and their detail is infinitely intricate. The closer you look, the more detail you see.

When they were first invented fractals were considered to be nothing more than weird mathematical oddities. However, in recent years scientists have discovered that fractal shapes are extremely common in nature. The shape of a cloud is best described as a fractal. The path boiling water follows as it seeps through the ground coffee in your coffee maker is the shape of a fractal. In fact fractal geometry is playing a vital role in the development of an exciting new frontier in physics — the study of complexity.

This new frontier has developed from recent breakthroughs in physics that have shown that some seemingly complex phenomena may in fact have exceedingly simple explanations. For example, it may be possible that the complicated motion of turbulent waterfalls could be explained by a few simple rules. The turbulent eddies themselves may show fractal behaviour.

Fractals come in many varieties. Just as there are different shapes in standard geometry there are different fractal shapes. Try imagining the following simple fractal.

Draw an equal-sided triangle on the middle of a page. Rub out the middle third of each side and replace each by a smaller equal-sided triangle. Then rub out the middle third of each side of each of the three smaller triangles and replace these by even smaller equal-sided triangles. Now imagine you could repeat these steps over and over again for ever! The object you would create is called the Koch curve. It is a simple fractal.

Just like a coastline, the outer edge of the Koch curve is infinitely long and the more you magnify it, the more detail you see. Another interesting feature of a fractal is that the pattern is the same at higher and higher magnifications. No matter what strength magnifying glass you use, the Koch curve always looks the same.

One very practical application of fractals is in the oil industry. To enhance recovery from an oil field, water is injected into the middle of the field. This 'bubble' of water helps drive the oil to the surface where it can be easily extracted. You might think the shape of this water bubble would just be an amorphous blob or, at best, a sphere. However, instead the water forms a beautiful and intricate pattern of long fingers and branches. If any of the fingers are magnified, even more intricate fingers and branches are revealed. The shape of the water bubble is a fractal.

The more science learns about fractals, the more we discover just how much Mother Nature seems to like them. They appear everywhere. The task now is to discover just why they occur.

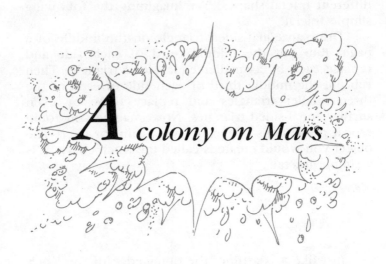

# A colony on Mars

*O*ne of the big obstacles preventing a human colony from living on Mars is that the Martian atmosphere is unbreathable. Rather than being rich in life-giving oxygen, its atmosphere is abundant in carbon dioxide. However, scientists are now giving serious consideration to ways of solving this problem.

One idea is to build a completely enclosed, Earth-like environment on Mars, an artificial ecosystem capable of sustaining human life. In fact the building of such structures, called biospheres, is already well under way. Recently an American marine biologist, Abigail Alling, spent five days in an artificially created ecosystem. The structure, which looked like a giant greenhouse, provided all the food, air and water Alling needed. Everything was generated and recycled within the artificial ecosystem.

The group behind the project is Space Biospheres Ventures and the five-day test is just the beginning. In September 1990 they are planning to put four men and four women into an artificial ecosystem for two years. The only thing that will enter their biosphere from outside will be sunlight. Everything else will be generated internally. If successful, this will be a major advance towards extraterrestrial living.

Mars has always had a certain mystique about it. In many ways it is very similar to our own planet. It has a twenty-four-hour day, drifting white clouds and even icecaps at the north and south poles. In fact the pictures sent back by the Viking spacecraft that landed on Mars in 1976 showed the Martian surface to be strikingly similar to a desert landscape on Earth. Apart from the pink sky, of course! People have dreamed of visiting this red planet and setting up colonies for years. In fact for many years many people believed that Mars already had life of its own.

Around the turn of this century, a man by the name of Percival Lowell carried out many detailed

observations of Mars. These observations were to become legendary. Lowell was convinced that, through his telescope, he could see a system of canals networked across the surface of the planet. He believed that the canals were constructed by intelligent beings to bring water from the polar caps to the arid parts near the equator. He spent many hours making detailed maps of these canal systems and even produced globes detailing the Martian surface.

However, as time passed, the Lowell legend faded. With the construction of bigger and better telescopes the Martian canals seemed to disappear. Even at the time Lowell made his startling observations, there were other astronomers who denied seeing them. Sadly, it seemed that the more we explored Mars, the more unlikely the existence of intelligent Martians.

However, the question of life on Mars is a different matter. It is quite possible that Mars is teeming with microscopic life, like that which inhabited the early Earth. In 1976 this possibility was investigated when two spacecraft, Viking 1 and Viking 2, landed on Mars to test it for life. Each set off with three different experiments designed to do the test. Unfortunately all the results were negative.

Although no signs of life were found, this didn't rule out life on Mars. Searching for life on other planets poses problems. To set up any test for life, certain assumptions about what that life is like must be made. And since the only life we know about is life on Earth this must colour the assumptions. The three experiments carried out on Mars in 1976 simply failed to detect any signs of the type of life we were *looking for*. But life detectors have been wrong in the past.

Before 1973, conventional tests used in Antarctica failed to detect any native microbial life. However, in November 1973 a microbiologist named

Wolf Vishniac tried a new test that he had developed and found an abundance of Antarctic microbial life. Unfortunately, Vishniac's test wasn't developed in time to be taken on board the Viking spacecraft. If it had, our understanding of living conditions on Mars may have been very different.

# How much can you blame on your parents?

*J*ust how much can you blame on your parents' genes? For some things, there's no doubt about it. Ugly noses, crooked eyes and even stumpy legs can often be traced back to either mother or father's genes. But what about behaviour? Can your bad tempers and violent outbursts be blamed on your parents' genetic make-up? Or is human behaviour something that is learned mainly from the environment? The answers to these questions are not easy. While the environment obviously plays a big part in shaping behaviour, your genetic make-up also plays a role.

Sex hormones, for example, seem to play some part in fostering aggressive behaviour. One study of United States prisoners found that violent prisoners had levels of testosterone almost twice as high as the normal population. Even studies of university wrestlers reveal that the winners often have higher levels of testosterone.

The effect of testosterone on aggressive behaviour has been known for a long time. The family tom-cat shows a distinct preference for the quieter life after a large part of his manhood has been removed. The castration, of course, reduces the cat's levels of testosterone. In fact farmers have routinely used castration as a means of calming belligerent male animals for centuries. So it should be no surprise that hormone levels in humans can affect their behaviour. It seems that some people may be more predisposed to violent behaviour than others.

However, human behaviour is greatly affected by the environment. Growing up with aggressive parents or living in an aggressive environment can lead to learned aggressive behaviour. There is even mounting evidence that simply watching violent videos can cause the viewers to become more violent.

There has always been strong debate over the importance of environment versus genes when it

comes to human behaviour. And a lot of money and scientific work have been directed towards trying to find answers. However, the debate goes much deeper than purely understanding human behaviour. It's a theme that occurs throughout biology. It is crucial to the question of how a living creature interacts with its environment. And one of the more important areas where answers are needed is in the study of diseases, in particular, the study of the disease that is the second greatest killer of Australians: cancer.

One of the things known about cancer is that it is often found to run in families. Now at first this may seem like strong evidence that genes are a very important factor for determining whether a cancer will develop. However, that's a premature assumption. Genes are not the only things shared by families. Families also often have similar eating, drinking, exercise and sexual habits. In fact studies seem to show that these environmental influences on cancer are more important than inherited influences.

The strongest evidence for this comes from studies of migrants. When people change countries they usually also experience a radical change in environment. And since a person's genes can't change, any variation in the type and rate of cancer must be due to the different environment. For example, in Japan the rates of breast and colon cancer are relatively low, while cancer of the oesophagus and stomach are high. However, the statistics for the Japanese who have emigrated to Hawaii are very different. In fact these Hawaiian Japanese actually acquire the patterns of cancer found in America, that is, relatively high levels of colon and breast cancer and lower levels of oesophageal and stomach cancer. Since Hawaiian Japanese have the same genes as the Japanese in Japan, the different cancer pattern must be due to the different environment.

One of the environmental causes of cancer is thought to be food. There has been a lot of publicity lately about the various chemical additives found in food and their possible effects on the body. However, studies conducted at places like CSIRO's Division of Human Nutrition seem to show that these are not the main factors causing cancer. It seems that a more important concern is the overall balance of nutrients in the diet.

The finely balanced chemical processes in the body's cells can be upset by eating the wrong balance of foods. And this upset balance can sometimes lead to cancer. For example, diets high in fat, salt and alcohol with too little dietary fibre seem to increase your risk of cancer. And vitamin deficiencies are also considered to be a problem.

For example, researchers believe that oesophageal and stomach cancers may be greater in less-developed countries because their diets are more likely to be deficient in specific nutrients such as vitamin A. And the higher consumption of pickled, salty and mouldy foods may add to the problem. In contrast, the diet in countries like Australia, which are rich in nutrients, particularly fat and protein, could be resulting in an increase in cancer of the bowel.

The effect of diet on cancer is an area that requires much more research. Many things are still not understood and even the things that are becoming clear require a lot more evidence. However, despite this, the current recommendations from CSIRO are fairly simple. Don't consistently eat foods high in fat and protein, and avoid obesity. Eat plenty of fresh vegetables and don't eat too much highly processed food. And, finally, only drink a moderate amount of alcohol. These recommendations are actually very similar to those made by the Heart Foundation for reducing your risk of heart attack. So even though all the evidence isn't in yet, the advice is to play it safe.

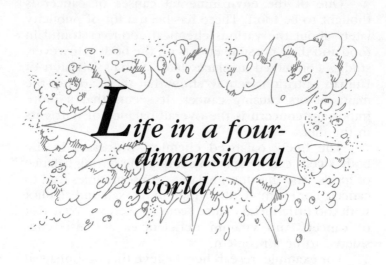

# *L*ife in a four-dimensional world

*R*ight in your very backyard, in broad daylight, you see a small spherical object hovering about two metres above the ground. You rush over to it. It's like a marble suspended in space. You run your hands all around it. There are no strings attached. Suddenly the marble starts growing, expanding like a perfectly round inflating balloon. It grows to the size of a giant beach ball and hovers around you in a circle. Then, as fast as it came, it shrinks to the size of a pin head and vanishes into thin air.

This is a description of what you might see if you were visited by a spherical spacecraft from a fourth dimension. If it wasn't spherical what you might see would be even more bizarre — a weird object dancing, expanding, changing shape and even splitting into two or more pieces. Of course we have never actually been visited by anything from a fourth dimension, but surprisingly it is possible to imagine what such a visit might be like. In fact mathematicians routinely work with four- and even higher dimensional spaces.

The world is said to be three-dimensional because three numbers are needed to specify the exact position of any object. For example, to specify the exact position of this book three numbers must be given. These could be its distance from the northern wall, the distance from the eastern wall and the distance above the floor, or, if you prefer, the longitude, latitude and height above the ground. After all, you may be reading this in an aeroplane. But no matter how you specify its position, you must always give at least three numbers. In the same way all objects are three-dimensional — they have length, width and height. However, although the real world is three-dimensional, it's not hard to imagine living in a two- or even one-dimensional world.

Divers working on underwater pipelines have a chance to experience what life might be like in a one-dimensional world. Before the widespread use of robots, it was the diver's job to swim along inside the pipe and fix any problems that occur. Some of these pipes are so narrow that the diver must unstrap his tank and swim through, dragging it behind him. Living inside a narrow pipe would be like living in a one-dimensional world. Only one distance, say the distance the diver has travelled along the pipe, is needed to know his position. He only has one dimension to move in — forward and backward. His ability to move up and down and left and right has been taken away.

If you feel the desire to experience what it would be like to live in a 2D world, try going under your house. If your place is anything like mine, you don't walk while under the house, you slide around on your stomach, bumping your head continually. Under the house you only have two dimensions to move in. The third one, up and down, has been taken away.

In a four-dimensional world four numbers would be needed to know the exact position of a 4D newspaper: the distance from the north wall, the distance from the eastern wall, the distance from the floor, plus one other — a number which measures the newspaper's position in the fourth dimension. All objects in a 4D world would have length, width, height and another dimension at right-angles to these three. Don't try to imagine it, it's impossible for us poor 3D humans.

But getting back to the original question. What would we see if a four-dimensional creature visited our 3D world? Because our minds can't imagine four dimensions you might think this is an impossible question to answer. However, the trick is to move down a dimension. What would we look like to a two-dimensional being if we invaded its world?

Try this experiment. Take a sheet of newspaper and cut a narrow slit window in it, one quarter of a centimetre wide and about thirty centimetres long. Hold the paper in front of you at arm's length with the slit held horizontally. Looking through the slit gives you some idea of what things look like in a 2D world. Your ability to see in the up and down direction has been taken away. Take a tennis ball and move it up and down behind the newspaper. All you see at any one time is an edge on view of a thin cross-section of the tennis ball. The 3D ball looks like a 2D disc through your 2D vision. As you move the ball up and down the disc grows and shrinks. If you move it up so far it leaves your view altogether, the disc appears to shrink and then vanish out of existence.

Move everything up one dimension and you have an explanation for what we would see if a four-dimensional spherical spacecraft visited our 3D world. If a four-dimensional tennis ball passed into our three-dimensional world all we would see at any one time is a 'three-dimensional' cross-section. A 3D cross-section of a 4D ball is a normal 3D ball.

Confused? Well, brace yourself for more. A 4D man would also have a type of X-ray vision. He would be able to see your insides in their entirety. Again, to imagine this move down a dimension. What would we 3D humans see if we looked into a 2D world? Imagine what life would be like if you lived inside an extremely thin piece of paper. You could move left and right and backward and forward but you could not move up and down since this would take you out of your world. In fact the direction up and down would simply not exist. Such a world is two-dimensional and only very flat beings could inhabit it. For example, such creatures might be small paper-thin circles. For the moment imagine the circular creatures are red near the centre and green near the edge. In other words, they have green skin and red insides. What would one circle see when he looked at one of his circular friends? All he could see is the green edge, or skin. There is no way he could see through the skin and look at his friend's red insides. Remember, it is impossible for the circles to leave their 2D world, inside the piece of paper.

But we are different. Because we are 3D we can look down on the circle people and see their green skin and red centre simultaneously! If we described a circle's insides to him he would be convinced we possessed some sort of X-ray vision. In the same way it would be possible for a 4D creature to see our skin and insides simultaneously by looking down on us from his fourth dimension.

Of course, none of this proves that 4D creatures exist, but it's nice to know what to look out for in case you meet one.

# *F*loating islands

*W*hile watching a glowing sunset over the west coast of the Falkland Islands it must be hard to imagine that several million years ago the sun used to set over the east coast! It's not that the sun has moved or even the Earth. It is the Falkland Islands themselves!

Research at Oxford University has found that over the last 200 million years the islands have rotated through 180 degrees. The south coast is now the north coast and the east coast is the west. Rather than facing South America, the west coast used to face Africa!

These days the idea of floating islands is nothing new. In fact it is well known that everything from the smallest island to the largest continent is adrift. While you are reading this book Australia is voyaging north-wards on a collision course with Asia. Not that there is any need to rush out and cancel that trip to Singapore — the drift speed is only a few centimetres per year.

Although this is slow, over millions of years the continents can move thousands of kilometres. And when they collide the results can be spectacular. When India crashed into Asia the land buckled to form the tallest mountain range on the face of the Earth. Even now India continues to burrow under Asia, pushing the Himalayas to even greater heights. In fact, since early man settled in the region, the Himalayas have grown about 1,500 metres!

Although this theory of continental drift is well accepted today, there was rather a different reaction when the idea was first suggested. A German scientist, Alfred Wegener, proposed the theory as early as 1912. However, his work gained little support and by the end of the 1920s it was officially deemed impossible. The path from almost universal rejection to complete acceptance is a fascinating tale.

Earlier this century scientists were puzzled by a certain fossil plant (*Glossopteris*) which had lived about 350 million years ago. The fossil evidence showed that these plants had flourished in Australia, South America, India, Antarctica and South Africa. The plants spread via tiny spores which were carried by the wind, animals and insects. The problem was that no matter how strong the wind or how hardy the insects and animals, how could the spores possibly be carried across the enormous stretches of ocean that separate the continents? The answer is simple: they couldn't be.

This enigma is one of the things that inspired Wegener's theory of continental drift. Spurred on by the snug fit between the east coast of South America and the west coast of Africa he suggested that at one time all the continents had been joined together. Like pieces of a jigsaw puzzle, they formed an enormous, single mass of land. He believed that at a time in the past some unknown cataclysm caused this supercontinent to break up. Each piece formed one of the present-day continents which have been adrift ever since.

If this theory was right, it was easy to explain how the *Glossopteris* had moved from continent to continent. It had simply spread before the continents separated.

However, the idea of drifting continents seemed ludicrous to geologists in the early twentieth century. Instead they believed that the continents must have once been connected by bridges of land that had since sunk into the ocean. This theory of sunken land bridges turned out to be very convenient. Not only did it mean that there was no need to believe in drifting continents but also this theory couldn't be tested. Geologists could sleep at night!

In fact they slept well for thirty years, until the invention of echo-sounding equipment capable of making detailed maps of the ocean floor. Using this equipment no signs of sunken land bridges were ever found. Suddenly, Wegener's theory was back in the ring. In fact, two new discoveries concerning the ocean floor seemed to support Wegener's ideas.

First of all, it had always been thought that the ocean floor was virtually flat. However, this idea was shattered with the discovery of a tremendous underwater mountain range beneath the Atlantic Ocean. The tallest peaks of this mountain range rise over 3,000 metres above the sea floor. This discovery led to further exploration and eventually a chain of underwater mountains encircling the globe was found.

The second discovery concerned the age of the ocean floor. It had always been believed that the ocean floor was almost as old as the Earth itself. However, measurements of the age of the Atlantic's floor showed that even the oldest parts were only a small fraction of the age of the Earth.

It was these two clues that led to the modern theory of continental drift. Continental drift can be easily understood by thinking of the Earth as a hard-boiled egg. Just as an egg has three different regions, the yolk, the white and the shell, so does the Earth. These are the core, the mantle and the crust respectively. However, unlike an eggshell, the Earth's crust is not uniform. Some parts are thicker than others. The continents are regions where the Earth's crust is thick enough to protrude out of the ocean while the thinnest crust forms the ocean floor.

To complete the model, drop the egg on the floor, cracking the shell into several pieces. If you can imagine that each piece of the shell was free to slide over the surface of the egg white, you would have a good understanding of continental drift.

The Earth's crust is actually cracked into several pieces which are sliding over the mantle. So it's not only the continents that are drifting across the Earth but the ocean floor as well. When two thick pieces of crust collide (that is, two continents collide) a mountain range is formed. When two pieces of crust separate, lava gushes up to fill the gap. As the lava comes into contact with the sea it hardens, forming an enormous volcano stretching the entire length of the crack. This is precisely how the mid-ocean mountains were formed. As the two pieces of crust continue to separate the gap is continually filled by new hardened lava. In this way new crust is continually created at the mid-ocean ridges.

However, just as there are places where crust is continually created, there are also places where it's destroyed. When thick and thin crust collide (or, in other words, when a continent collides with an ocean floor) the thick continental crust rides over the thin oceanic crust, pushing it down into the Earth's hot interior. Here it is melted down again. In this way the thin parts of the Earth's crust which form the ocean floor are continually recycled.

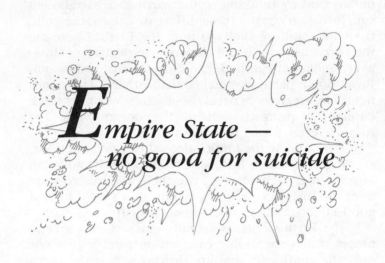

# *E*mpire State —
## *no good for suicide*

*D*uring the Christmas of 1977 John Helms, a lonely artist, learnt the meaning of the word luck. Forlorn and deeply depressed, he found himself clinging to the safety fence of the eighty-sixth floor of the Empire State building. After a brief prayer he flung himself towards his death. Half an hour later, however, he regained consciousness. Instead of landing at heaven's doorstep he had landed on a one-metre-wide ledge on the eighty-fifth floor! A gale-force wind had blown him to safety.

If it wasn't for the wind Helms would have hit the pavement at a blistering 190 km per hour! A gruesome detail you could no doubt do without. However, in recent years, the speed of falling bodies has become a matter of hot debate. It started back in 1986 with the publication of some results that seemed to show that Galileo's legendary cannonball-dropping experiments were wrong. Legend has it that Galileo dropped a musket ball and a cannonball from the

Leaning Tower of Pisa to demonstrate that both would fall at the same rate and hit the ground together. However, the work published in 1986 by Ephraim Fischbach and his colleagues seemed to show that Galileo's conclusion was incorrect. Rather, it seemed that the lighter object, the musket ball, would fall faster and therefore hit the ground first! The scientists believed this was due to an as-yet-undiscovered force in nature, a force that acts in opposition to gravity.

Since then there have been bitter debate and many new developments. However, before continuing, a crash course in the physics of falling bodies would be helpful. If you had to choose between jumping from the top of the Empire State or an aeroplane, which would you take? Surprisingly, it makes little difference. Whether you jumped from an aeroplane, or an orbiting satellite for that matter, you would still only hit the ground at around 190 km per hour!

The reason there is a maximum speed of fall is due to air resistance. At about 190 km per hour the force of air resistance just cancels the force of gravity. With no force there can be no acceleration. And with no acceleration there can be no further increase in speed. So once having reached terminal velocity, whether you fall another metre or another million metres is irrelevant.

Of course jumping from a satellite could pose problems when re-entering the Earth's atmosphere. If suitable precautions weren't taken you would shine briefly as a falling star before burning to a cinder.

Now air resistance can play some devious tricks. In fact if you want to win a few dollars at the pub try this question: if a beer belly champ and a jockey jumped together from an aeroplane, which one would hit the ground first? The most common answer is the beer champ, because he is much heavier. However, it's a common misconception that heavy bodies always fall faster than light ones. It's not weight that determines the rate of fall, it's air resistance.

I've been using 190 km per hour for the speed of falling humans but this is by no means a magic number for all falling objects. It varies from object to object and person to person, and even depends on how the person is falling. A skydiver falls belly first with head and feet arched backwards at approximately 190 km per hour. However, by moving into a vertical dive position, terminal velocity can be increased to more than 320 km per hour! So depending on how each falls, either the jockey or the beer champ could hit the ground first.

Although there is no general rule for the speed at which an object falls through air, there is a very definite rule for objects falling in a vacuum. Galileo stated that in the absence of air (on the Moon, say) all objects, no matter what their size, shape or weight, fall at the same speed. On the Moon a feather and a brick fall at equal speeds. If released together from a height of 3,000 metres, for example, the brick and feather fall side by side all the way and eventually crash into the ground at 360 km per hour! Even a feather could do damage travelling at this speed. Because there is no air resistance on the Moon there is nothing to balance gravity. As a result all falling objects are accelerated to ever-higher speeds.

Galileo announced his law in the seventeenth century and it went unchallenged until 6 January 1986 when Fischbach and his colleagues published their results. According to their work, on the Moon the feather would hit the ground before the brick.

However, it now seems that Fischbach and his colleagues may have been mistaken. In a 1988 meeting of physicists the consensus opinion was that there is no firm evidence for a new force that causes lighter objects to fall faster than heavier ones. Ironically, however, work by other groups trying to validate Fischbach's results have found evidence for not one but two new forces. Unlike Fischbach's force, these forces seem to affect all objects equally, one of them opposing gravity and the other acting in the same direction as gravity.

A number of groups around the world are now conducting research in order to determine the existence of these new forces. Experiments involving the measurement of the gravitational force in mineshafts and in boreholes in Greenland ice look for deviations from the conventional gravitational laws. Some experimenters have even made modern-day equivalents of Galileo's experiment. For this, small cubes of uranium and copper are dropped in evacuated chambers.

Of course the research in this area is still at a very early stage. It is still too early to state with any confidence that extra forces are present. Hopefully, during the next few years the evidence will mount one way or the other.

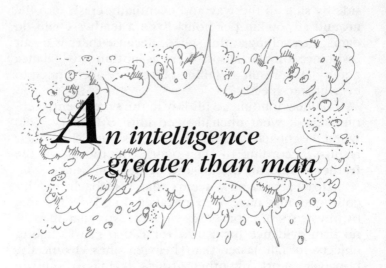

# *A*n intelligence greater than man

*W*ill we ever meet an intelligence greater than ours? A creature with an intelligence and understanding of the world way beyond our comprehension? Most UFO enthusiasts would say there is little doubt about it. Not only does intelligent life exist elsewhere in the universe, but planet Earth is already being visited by it. Most scientists, however, do not agree. While they do agree that it is highly likely that intelligent life does exist elsewhere in the universe, they say that it is extremely unlikely that we will ever come into contact with it. The problem is that even if there were many planets teeming with intelligent life, the distances between them would most probably be so enormous that the likelihood of one of our extraterrestrial neighbours stumbling across Earth is extremely remote.

However, despite this, many scientists still believe that we will one day meet an intelligence greater than the human mind. Rather than from space, this

**144**

intelligence will come from Earth. It will be a computer.

Despite the feelings of inadequacy they create among many users, present-day computers are really nothing more than sophisticated machines. However, if there is one thing more awe-inspiring than the capabilities of a computer, it is the rate at which they are being improved. In only a few decades they have already mastered many of man's abilities and in many areas their performance well outstrips that of man.

Is there any limit to the possibilities for a computer? Will the time come when we construct a computer that is superior to man in everything he does? A machine that can ask questions, have feelings and be genuinely creative? Many computer scientists believe such a machine will eventually be developed and they say it will be very 'human'. The possibility raises some profound philosophical questions. Would such a machine be conscious? Should an artificial intelligence

be given equal rights to a human or perhaps even greater rights?

Deeply perplexed by such questions, Alan Turing, a mathematician, developed a test for an artificial intelligence. The Turing test is intended to be an objective way of determining whether a computer can 'think'. It requires a human subject, a human tester and the computer to be tested. The idea is that the computer and the human subject are both hidden from the tester. The tester then asks the human and the computer a series of questions and, from the answers alone, must determine which one of the two is the computer. If the tester cannot determine any difference between the human subject's and computer's responses, the computer is said to have passed the test. The argument is that, if no one can tell the difference between the computer and a human, there is no choice but to assume that the computer is thinking.

The area of science concerned with developing computers with the capabilities of the human brain is called artificial intelligence. Many scientists working in artificial intelligence believe that it is only a matter of time before we develop a computer capable of passing the Turing test. Many also believe that once it has done so, terms such as thought, understanding, awareness, happiness, pain, compassion and pride could be applied to such a computer.

This opinion has come to be known as the 'strong AI (artificial intelligence) point of view'. Other computer scientists take a less strong stance. In particular, John Searle, a philosopher from the University of California, believes that although we may develop a computer capable of passing a Turing test, it does not automatically follow that this computer can 'think'. He explains this in terms of an example.

Imagine a person, who does not speak a word of Chinese, in a room with a box of Chinese symbols and a book. The book contains a list of questions and their answers and is written entirely in Chinese. Outside the room is a tester who speaks Chinese fluently and who also has a box of Chinese symbols. The task for the tester is to determine whether the person in the room can speak Chinese. He does this by asking him a series of questions in Chinese. Both the tester and the man communicate with each other by passing Chinese symbols through a window.

The task of the man inside is to answer these questions and fool the tester into thinking that he actually does understand Chinese. He does this using the book. He simply looks up each Chinese question in the book and finds the Chinese answer. Providing that the book contains every possible question the tester might ask, his answers would be perfect. The tester would be convinced that the man in the room understood Chinese. Of course the man in the room hasn't understood a single question. Instead he has simply repeated the answer beside each question in the book.

John Searle argues that precisely the same thing can be said about a computer capable of passing the Turing test. Just because its responses are indistinguishable from those of a human doesn't mean that it necessarily 'understands'. It may still be nothing more than a very sophisticated machine.

# *Immortality*

Anyone who's serious about immortality must have looked into cryonics. Cryonics is the preservation of human corpses by freezing. The idea is that upon death, your body is snap-frozen and stored in the deep freeze until the time comes when science has discovered a way of resurrecting you. By then, its patrons hope, science will have also discovered a cure to the disease that killed them and they will be free to take up their lives where they left them. In fact for a fee, a considerable fee, there are now companies around that offer this cryonic service.

Some customers place even more faith in science. As a cost-cutting measure, they ask to be decapitated upon their death and request that only their heads be frozen. Their hope is that not only will science have discovered a way of bringing them back, it will also have discovered how to give them a new body.

Now anyone who has eaten defrosted straw-berries knows that freezing can have a dramatic effect on biological material. And although there has been much research into the effects of freezing on living tissue, cryobiology, as the subject is called, is only in its infancy. The big challenges at present are to discover ways of freezing and defrosting individual organs such as kidneys and livers. Unfortunately, applying the technique to complete human bodies is a long way off. Successful cryonics still remains very much in the pages of science fiction.

Those who wish to obtain immortality through cryonics must make two acts of scientific faith. First they must believe that science will eventually come up with a way of successfully thawing them and second they must also believe that the secrets of immortality will one day be discovered.

For the first of these, things don't look promising. The problem with freezing living tissue is that ice crystals wreak havoc on cells and blood vessels. While it is true that some animals can survive subzero

temperatures, their secret is never to become completely frozen. For instance, certain species of frogs can survive temperatures as low as − 8 degrees. However, even at this temperature, not all of their body water is frozen.

However, for the second act of faith, some like to have hope. In principle at least, it is possible to conceive mankind one day achieving something approaching immortality. As the surgeon becomes more skilled with the scalpel, as more diseases are cured and as the problems of old age solved, life expectancy will soar. In fact during the last century or two, we have seen science already make massive advances along this path. Many of the diseases that were once great killers are now under control and the advances in surgical techniques have been enormous. In fact life expectancy, in western nations at least, has never been higher. However, the future offers even greater medical breakthroughs. In fact medical science is already entering a dramatic new era — the era of genetic engineering.

Using genetic engineering, scientists can make changes to a living creature by actually altering its genetic structure. An animal's genetic structure contains a tremendous amount of information on how to perform every instinctive action, such as breathing, digesting food and combating disease. In fact an animal's genes literally store a small library of information. If all the information contained in the human genetic structure was written down, for example, it would fill a hundred thick volumes. However, surprisingly perhaps, all of this genetic information is encoded on to a single molecule, the master molecule of life, DNA. It is the DNA molecule that issues the instructions for running the body.

Like all molecules, DNA is built from atoms chemically bonded to other atoms. And its structure can be changed by simply changing these chemical bonds. By changing the structure of the DNA molecule, the instructions it issues to the body also change. For example, some people of African descent have red blood cells that are shaped like the crescent moon, rather than the more usual globular shape. These sickle cells provide a major resistance against malaria, although they also transmit a type of anaemia. However, if you live in malaria country, it's better to be anaemic than dead. This rather major difference between these Africans and Europeans is due to an extremely small difference in their DNA molecules.

Genetic engineering is making massive in-roads into all the life sciences. It's used to create new types of bacteria, new plants and even new animals. However, one of its most exciting applications is medicine. Already it is used in the search for new treatments and vaccines for disease, but in the future it may be used for even greater things, such as gene therapy. Gene therapy involves replacing damaged or faulty genes with new ones. Many diseases result from genetic defects. Diseases such as haemophilia and cystic fibrosis, for example, are effectively programmed into the sufferer's DNA. If the faulty genes responsible could be replaced, these diseases could be eradicated.

Many experiments on gene therapy have already been conducted on animals but recently the first approved tests for putting new genes into a human subject commenced. Genetically altered cells were injected into a patient suffering from a malignant skin cancer in its advanced stages. Many eyes will be watching this pioneering work. If gene therapy for humans becomes commonplace and if genetic engineering can lead to the eradication of the two greatest killers, heart disease and cancer, life expectancy will sky rocket. Human immortality will move one giant step closer.

# *R*unning on sunshine

*F*irst it was cameras to catch you racing the red light and then later came the speed camera. Now the eyes of big brother have gone even further. The humble parking attendant now has access to the latest in spying equipment. The parking meter has gone high-tech. In the Victorian town of Shepparton, twelve new parking meters have recently been tested.

The meters will not only perform the tasks of conventional parking meters but will also keep track of the cars as they arrive and depart parking spaces and will instantly alert parking officers if money is not put in the meter. The new high-tech meters even cancel the remaining time when a car departs. However, although the meters are high-tech, they are powered by one of the oldest sources of energy, solar power.

Solar power is slowly creeping into many areas of modern life. There are solar-powered refrigerators, solar-powered cars, solar-powered calculators and hot-water services. There are even solar-powered spacecraft. However, despite these, solar energy still only forms a very small percentage of the world's total energy consumption. By far the majority of the world's energy comes from burning fossil fuels like coal, oil and gas. And unfortunately it seems that the world is beginning to suffer some of the consequences of using this cheap and easy fuel. The greenhouse effect has finally made front-page news. The slow atmospheric build-up of carbon dioxide, a by-product of burning fossil fuels, seems set to cause a worldwide warming that could mean disaster for many.

At a time when concern for the atmosphere is mounting, solar energy offers a clean and safe alternative. And there is certainly no shortage of sunlight. The power reaching the Earth's surface from the Sun is about 10,000 times greater than the present worldwide power consumption. The problem, of course, is converting this solar energy into useful forms of power. Using current technology, solar cells can be made that are about thirty per cent efficient. This means that a thousand watts of sunshine can be converted into 300 watts of useful power. So in terms of energy, if enough solar cells were made there would be more than enough energy to satisfy current world demands. Given our environmental problems you might wonder why we haven't made more use of solar power.

Of course, like most things, it comes down to a question of economics. Although the sunlight is free, the solar cells needed to convert it into useful energy are not. It turns out that for many applications, energy from fossil fuels is cheaper than solar energy. One exception, however, is power for remote locations. It is often very expensive to take conventional power to very remote places. Remote lighthouses, radio repeaters and on board spacecraft are ideal applications for solar power.

A solar cell converts sunlight into electrical energy. This is achieved by a layer of material such as silicon or gallium arsenide that can absorb the incoming sunlight. As the sunlight is absorbed, a positive and negative charge are created in the material. These charges then move to different regions of the material and a voltage is created across the solar cell. This voltage can then be used to drive an electrical current.

Although the use of solar power has so far been restricted by economics, the greenhouse effect has added a new term in the economic equation. What price can be put on the environment? If we are to avoid the greenhouse effect, in the future more use must be made of alternative energy sources. Serious consideration must be given to energy from the Sun, the waves and the wind.

Research into wind power is already taken quite seriously in many parts of the world. In the Netherlands and the United States, for example, huge fields filled with windmills have been set up to extract useful power from the wind. Once again, however, for most applications wind power has proved to be uneconomical. It has only shown itself to be a real alternative in remote, windy areas.

Nuclear power is seen by many as the best alternative energy for the future. And although nuclear accidents and nuclear waste pose major problems, nuclear power does at least keep the atmosphere clean. However, the world hasn't caught on to nuclear power as quickly as many predicted. In fact the OEDC has had to revise its projections for the future use of nuclear power consistently downwards each year. And the nuclear accident at Chernobyl has no doubt added to the problems.

However, the supporters of nuclear power hope that the nuclear energy of the future will be very different from that of the present. The current generation of nuclear reactors gain their energy from nuclear fission, the process of splitting atoms. However, nuclear energy is also released when two atoms are fused together. This process, called nuclear fusion, is the way the Sun derives its energy and is also the process behind the hydrogen bomb. The aim of fusion research is to produce a controlled nuclear fusion reaction that could be used in a reactor. If successful, mankind will be guaranteed a relatively safe and clean source of energy that is almost inexhaustible. The problems of radioactive waste from nuclear fusion are minimal.

# Creating life with a chemistry set

*D*uring the 1950s, Stanley Miller tried an unusual experiment. In a reaction vessel, he mixed five very simple chemical substances — hydrogen, water, methane, ammonia and hydrogen sulphide. He then subjected the gaseous mixture to a series of electrical sparks. Since each of the gases in the mixture is transparent, the reaction vessel initially appeared empty. However, after ten minutes of sparking, a sticky brown goo started to form on the sides of the reaction vessel. When the experiment was completed, the brown tar was taken out and examined. It was revealed that the tar contained a whole host of complex organic molecules. Among them were some of the very building blocks of life.

The purpose of the experiment was to recreate the conditions on Earth billions of years ago, at a time not long after the Earth had formed. The electrical discharges were used to simulate lightning, while the chemicals in the mixture were believed to be those present in the Earth's early atmosphere. Although the experiment was very simple, the results were remarkable. They showed just how easy it is to turn a few simple molecules into the building blocks of life.

Of course, making life's building blocks is a far cry from creating life itself. However, the experiments did lend support to the theory that life can be, and was, created from a concoction of very simple chemicals. In the real world, the brown tar was just the beginning. As time went on, ever more complex molecules developed. Finally, quite by accident, a molecule arose that was capable of making copies of itself. Using simpler molecules as parts, these molecules were the first thing on Earth that could reproduce. They set the stage for life.

Soon, other self-copying molecules arose, and the competition was on. Just as the competition for food among animals determines the path of evolution,

competition for molecular parts shaped these early chemicals. Only the fittest, or best adapted, survived. To compete more efficiently molecules with specialised functions developed and it was only a matter of time before these specialised molecules got together and formed the first living cell.

Further evolution saw the development of multicellular creatures. At first they were very simple and rudimentary, but with the passage of time ever more complex creatures evolved. Eventually, under the pressures of evolution, this early life evolved into every living creature on the Earth today.

However, just as life evolved, so did its environment. The Earth's primitive atmosphere was rich in hydrogen, which although perfect for the creation of life, would be highly poisonous for most living organisms on Earth today. As life evolved this early atmosphere was transformed into one containing life-giving oxygen and nitrogen. It was life itself that actually brought about the change. The story is best told from the beginning.

The Earth was born about four and a half billion years ago. Along with the Sun and all the planets, it condensed out of a giant cloud of dust and gas. As a result of its own gravity, the cloud collapsed, slowly compressing this gas and dust. The cloud literally collapsed under its own weight. To cut a long story very short, the result was the formation of the Sun, the Earth and all the other planets.

In its early days, planet Earth was a ball of hot molten rock. It was a world completely devoid of land, oceans and life. However, as the Earth slowly cooled, things began to change. The crust started to harden as temperatures on the surface lowered. And once the surface temperature dropped below the boiling point of water — 100°C — rain began to fall. Before long the first oceans had formed and the environment was ready for life.

The complex organic molecules created in the atmosphere dissolved in the newly formed oceans, forming an enormous organic soup. It was in this primordial soup that molecules got together to create life. Judging from the fossil record, it seems to have taken Nature about one and a half billion years to turn chemicals into multicellular creatures. Which leaves about three billion years for her to transform these earliest of animals into complex beings like ourselves. Complex life, however, is actually a relatively recent invention. For most of the three billion years the dominant life form was blue-green algae. However, although only a simple form of life, blue-green algae played an important role in the evolution of the environment. It prepared the atmosphere for life.

Blue-green algae produce oxygen as a waste product. As a result, the Earth's hydrogen-rich atmosphere was slowly replaced by one containing oxygen. Some of this oxygen reacted to form ozone, creating a protective layer high in the Earth's atmosphere. The ozone layer filters out most of the Sun's UV radiation, which although important for the creation of life, is very damaging once life has formed. Without the ozone layer life could never have ventured on to land. In fact, as we are becoming increasingly aware, an intact ozone layer is still vital to our existence on Earth.

Cast into the broader picture of the evolution of the Earth itself, the creation of life seems a very natural process. Is Earth the only planet where chemicals have been transformed into life? Since there must be countless other planets out there, all with lightning, UV radiation and the basic chemistry set required for life, extraterrestrial creatures seem highly likely. While some may not be as developed as us, others may be much further advanced. The question is, will we ever meet any of these extraterrestrials?

# The final winter

*N*o one really knows why the last dinosaur vanished from the face of the Earth about sixty-five million years ago. Over the years there have been many theories and no doubt many arguments. However, there is one especially interesting theory, interesting because there is now a very real possibility that the human race could become extinct for almost exactly the same reason.

In June 1980 a group of American scientists published a paper outlining new evidence supporting a radical theory for the extinction of the dinosaurs. They argued that a single event occurred about sixty-five million years ago which caused enormous amounts of dust to rise high into the Earth's atmosphere. As a result, light from the Sun was blocked for many months, plunging the entire planet into a prolonged savage winter. Without sunlight many of the plants would die. And with nothing to eat, a host of animals would soon follow suit. The scientists, who included Nobel Prize winner Luis Alvarez, proposed that this single event was the collision between the Earth and an asteroid. The result was the extinction of many creatures including the dinosaurs.

Their evidence came from a geological discovery. From an examination of sediments laid down about sixty-five million years ago it seemed that there was an unusually large amount of the element iridium on Earth at this time. Since asteroids contain considerably more iridium than the Earth's crust, the group reasoned that the extra iridium resulted from the collision between an asteroid and the Earth.

Although the theory was radical when it was proposed, it has slowly gained support in the years since. In fact at a recent meeting on global catastrophies, many of the delegates backed Alvarez's collision theory.

Around the mid 1980s, however, the mass extinction debate took a slightly different turn. A number of scientists suggested that the human race might also go the same way as the dinosaurs. Their reasoning was that nuclear bombs put up a lot of dust and debris high into the atmosphere. As a result a full-scale nuclear war might trigger another mass extinction, which would be caused not by radiation but by a long brutal winter. This time extinction would include the human race. This effect became known as the nuclear winter.

If there are any consolations of a nuclear war one must be that the bombing would probably be over very quickly. However, the devastation would be horrendous. The blasts from the bombs, the tremendous heat generated during the explosions and nuclear radiation would extinguish many people and cities. And on top of all this there would be delayed nuclear fallout raining back to the ground during the weeks that follow. A World Health Organization study estimated that about half of the world's population could perish due to these immediate effects. But the remaining half shouldn't feel too lucky. They would experience the nuclear winter.

During a nuclear winter temperatures could plunge well below zero over most of the globe and would stay that way for many months. Daytime light levels could fall to a daylight gloom or worse. The severity of the nuclear winter depends very much on where the bombs are dropped. If the targets are cities, for example, colossal fires would result and would fill the sky from horizon to horizon with black sooty smoke. Since soot is very efficient at blocking sunlight, temperatures on the ground would plunge very quickly. Large fires might burn uncontrollably for many weeks, especially since most of the fire fighting force would be out of action. As a result freezing temperatures would be maintained by the continual supply of obscuring smoke.

Fortunately, however, smoke particles are washed out of the sky by rain. So once the fires have stopped burning the atmosphere could quickly remove the smoke. Anyone who has been caught in the rain after a bush fire will have experienced the air-cleaning process. White clothes are quickly turned a murky grey.

The climate would recover quickly if smoke was the only factor. The problem is that rain clouds only form up to about fifteen kilometres above the ground. The fireball from a large nuclear blast is capable of drawing dust and debris to much greater heights. With no rain to wash it from the atmosphere, the high flying dust hangs around for years. And although dust is not as efficient as soot for blocking out sunlight it can still have a drastic cooling effect, which would last much longer than anything produced by smoke.

It's the combination of high-level dust with low-level smoke that would produce a nuclear winter. The amount of high-level dust depends on the size of the nuclear blasts. The amount of smoke depends on whether the bombs were dropped on cities or on military installations away from cities. Generally, away from cities there is nothing much to burn and smoke would be minimal.

Of course, it would be impossible to predict how many weapons would be used in a nuclear confrontation or where they would be dropped. Thus it is impossible to predict the length and severity of a nuclear winter. Even the calculations themselves contain many uncertainties and estimates. How much smoke would be produced when a city burns, for example? How fast would it be washed out of the air? However, in spite of these uncertainties, many scientists believe that the possibility that the entire human race will become extinct during a nuclear winter cannot be ruled out.

# Scientists make contact with Little Green Men

*I*n 1968 the *National Enquirer*, an American newspaper, reported that a group of scientists had made contact with another civilisation in space. Astronomers at the Cambridge University Observatory, the report read, had tapped into messages from an extraterrestrial intelligence. Strange as it may seem, the report did actually have some basis.

The story started with two astronomers, Anthony Hewish and Jocelyn Bell, who were using the Cambridge observatory to conduct what they thought would be some relatively routine observations. However, their routine night soon became exciting with the accidental discovery of an unusual set of radio signals. They discovered a very regular and continuous series of radio bleeps emerging from a certain place in the sky. The radio signal was unlike anything ever observed before. Hewish and Bell became excited. What sort of astronomical object could possibly cause such an orderly signal? Their imaginations began to wander. Could it be that they had stumbled upon an interstellar radio beacon used by an extraterrestrial intelligence? Or perhaps even a message from a distant civilisation. The two jokingly named the source of the bleeps LGM, for Little Green Men.

Hewish later admitted that these were the most exciting moments of his life. However, despite the excitement and the potential significance of their discovery, Hewish and Bell felt compelled to keep their work secret until the source of the radio signals could be explained. But keeping a discovery like theirs under wraps was no easy feat. Soon rumours were rife and eventually the *National Enquirer* got wind of the find. However, the *National Enquirer*'s premature report turned out to be a case of mistaken identity.

Soon after Hewish and Bell's discovery, other LGMs were found in different parts of the sky. Since it seemed unlikely that there were many civilisations out there all calling us at different frequencies the LGM

theory was soon abandoned. The objects were re-named pulsars and astronomers turned to an astro-nomical explanation for the regular radio pulses. Strangely, a hint of the explanation can be found in the history books.

On 4 July 1054 a Chinese astronomer noticed a brilliant star in the eastern sky that, until that night, had never been there before. It appeared in the con-stellation of Taurus the bull and was so bright that it could easily be seen during the day. The appearance of this new star was also witnessed by other early astron-omers around the world. Similar accounts are found in Japanese chronicles and there is even an Indian rock drawing in northern Arizona showing what is believed to be the appearance of the same mystery star.

After about 650 nights, the new star finally faded from view, never to be seen again. However, if a tele-scope is pointed to this part of the sky today one of the most beautiful objects in the universe can be seen. The Crab Nebula, a rapidly expanding red and white web of gas, is the gaseous remains of the explosion of a dying star. What the Chinese saw in 1054 was not the birth of a star but a star's death. The Crab Nebula is also one of the places where an LGM, or pulsar, can be found.

As a star comes to the end of its store of nuclear fuel it also comes to the end of its life. With no fuel to burn it can no longer provide enough thermal pres-sure to hold back its own gravitational force and the star literally starts to collapse under its own weight. If the star is big enough, the collapse will be violent and a tremendous explosion will result. This explosion, called a supernova, is so enormous that it can shine more brightly than a billion stars put together.

On average, about one in every thousand stars ends its life with a supernova explosion. This means that on average, we expect to see one in our galaxy about every hundred years. Since the last one occurred almost 400 years ago, in 1604, a supernova in the Milky Way is long overdue. In February 1987 the world did witness a relatively nearby supernova. It occurred in a neighbouring galaxy, the Large Magellanic Cloud. It was bright enough to be seen at night with the naked eye for many weeks before finally fading from view. A supernova in our own galaxy, however, would probably be bright enough to be seen even during the day.

After a dying star has exploded it continues to collapse packing its matter into an ever-smaller space. If conditions are right, the star eventually comes to rest as an incredibly dense ball of matter, a neutron star. A neutron star's matter is so compressed that one teaspoonful would weigh as much as a mountain. It turns out that, rather than little green men, a rapidly spinning neutron star can explain the regular set of bleeps from a pulsar.

Neutron stars have incredibly strong magnetic fields able to trap charged particles and cause them to emit a narrow beam of radiation. As the neutron star spins, this beam of radiation sweeps out across the universe, like a lighthouse beam across the ocean. Each time the beam sweeps past Earth we see a bleep. Since thirty bleeps are received every second from the Crab pulsar, it must be rotating thirty times each second. Considering neutron stars are about twenty kilometres across, this means that their rotation speeds must be enormous.

# Bibliography and further reading

Blakemore, C. 1988, *The Mind Machine*, BBC Books, London.

Blakemore, C. & Greenfield, S. (eds) 1987, *Mindwaves*, Basil Blackwell, Oxford, U.K.

Davies, P. 1987, *The Cosmic Blueprint*, Heinemann, London.

Gamow, G. 1965, *Mr Tompkins in Paperback*, Cambridge University Press, Cambridge, U.K.

Hawking, S. 1988, *A Brief History of Time*, Bantam, New York.

Hoyle, F. & Wickramasinghe, C. 1988, *Cosmic Life Force*, J. M. Dent & Sons, London.

Sagan, C. 1981, *Cosmos*, Macdonald, London.

Scientific American (eds) 1978, *Evolution: A Scientific American Book*, W. H. Freeman, San Francisco.

Silk, J. 1980, *The Big Bang: The Creation and Evolution of the Universe*, W. H. Freeman, San Francisco.